Orchid Biotechnology

ORCHID BIOTECHNOLOGY

edited by

Wen-Huei Chen

National University of Kaohsiung, Taiwan

Hong-Hwa Chen

National Cheng Kung University, Taiwan

World Scientific

NEW JERSEY • LONDON • SINGAPORE • BEIJING • SHANGHAI • HONG KONG • TAIPEI • CHENNAI

Published by

World Scientific Publishing Co. Pte. Ltd.

5 Toh Tuck Link, Singapore 596224

USA office: 27 Warren Street, Suite 401-402, Hackensack, NJ 07601

UK office: 57 Shelton Street, Covent Garden, London WC2H 9HE

Library of Congress Cataloging-in-Publication Data
Orchid biotechnology / editors, Hong-Hwa Chen, W.H. Chen.
p. cm.
Includes bibliographical references and index.
ISBN-13: 978-981-270-619-5 (hardcover : alk. paper)
ISBN-10: 981-270-619-4 (hardcover : alk. paper)
1. Orchids--Biotechnology. I. Chen, Hong-Hwa. II. Chen, W. H. (Wen Huei)
QK495.O64O55 2007
635.9'344--dc22

2007016772

British Library Cataloguing-in-Publication Data
A catalogue record for this book is available from the British Library.

First published 2007 (Hardcover)
Reprinted 2016 (in paperback edition)
ISBN 978-981-3203-41-9

Typeset by Stallion Press
Email: enquiries@stallionpress.com

Foreword

The *Phalaenopsis* is the national flower of Taiwan, first found and collected by the Japanese on Lanyu (Orchid Island) in 1897. After winning back-to-back championships in the International Orchid Exhibition in California in 1952 and 1953, Taiwan's native *Phalaenopsis* has gained worldwide admiration and the pride of the Taiwanese people.

With an optimal climate for growing the *Phalaenopsis*, Taiwan is well situated for commercialization of the flower, being in the northernmost area of *Phalaenopsis'* natural germplasm. Today, the *Phalaenopsis* industry in Taiwan is well developed and has advanced to green house breeding and systematic production. Moreover, Taiwanese product varieties now account for more than 50% of the global *Phalaenopsis* market share. It is no surprise that the *Phalaenopsis* industry is viewed as an example of the most advanced knowledge-based agriculture in Taiwan.

The future of the *Phalaenopsis* is exceptionally bright. On August 24, 2004, the New York Times reported that there is a $2 billion global market for orchids, with *Phalaenopsis* holding the leading share within that market. *Phalaenopsis* is also recognized as one of the most exciting and elegant indoor flowers by the American Orchid Society (AOS). And recently, Mr. Ed Matsui, owner of Matsui Nursery, the largest *Phalaenopsis* producer in the US, has estimated there will be a five-fold increase in the *Phalaenopsis* market within the next ten years.

Furthermore, according to the December 2004 issue of FloraCulture International, the most influential floral magazine, the *Phalaenopsis*, a newly developing flower with 20% growth each year for the past five years, is the top seller among all pot flowers in The Netherlands and Japan. Moreover, with the advent of mass retail as a new distribution channel, demand has increased for mini-type, low-priced product, and color variants, boosting flower sales in recent years.

All these factors have led to the *Phalaenopsis* being selected as one of the top four most important export products for Taiwan by the Agriculture Product Competition Module (APCM), a group developed by A-Turn Biotech Company, an advisor to the government for agriculture, to analyze and evaluate potential floral products for export.

However, the value created by the *Phalaenopsis* is greater than that of just the plant itself; strategic alliances with other industries will provide the opportunity for extracting further value and greater margin from the flower. *Phalaenopsis* is recognized as a symbol of elegance amongst flowers. Properly managed, this rare property enables the attraction and development of many complementary products and industries such as gifts, arts, and home decoration.

As Taiwan moves into the future, the establishment of acclimation and overseas sales points is one of the important steps in broadening the market of *Phalaenopsis* globally. Taiwan has a complete range of technologies from seedling acclimation to product vernalization. Through international strategic alliances enabling joint ventures and technology transfer, Taiwan will create higher profits through widening markets internationally for all parties involved.

Currently, due to shipping costs and importation regulations, certain finished *Phalaenopsis* products are not able to reach the US markets. Thus, establishing a local acclimation facility for consumers is now the most cost effective way to market and distribute *Phalaenopsis*. Fortunately, the US government has recently begun accepting the importation of *Phalaenopsis* with moss as a supporting medium, which will provide a new international trading opportunity for Taiwan.

There are many factors that contribute to Taiwan's unique capability to take the greatest advantage of the *Phalaenopsis* phenomenon. As mentioned above, the subtropical climate makes Taiwan a near perfect environment for the production of *Phalaenopsis*. In addition, Taiwan is rich in orchid species, holding a worldwide leading position in new product development fueled by hundreds of professional and amateur breeders who have won gold medals in world competitions. Taiwan has maintained and will continue to maintain its competitive edge on research and development in the *Phalaenopsis* business.

From a technical perspective, Taiwan's *Phalaenopsis* industry has the strength to compete with any of the major floral countries. It is strongly supported by The National Science and Technology Program for Agricultural Biotechnology (NSTP.AB), a joint program of the National Science Council, Council of Agriculture and Academia Sinica. The NSTP.AB supports universities and research institutions in advancing the technology in genetic transforming, tissue culture, and production.

Most of the chapters in this book have come about as a result of this National Project and I am honored to have served as the team leader since the initiation of the project in 1998.

For many years, significant advances in the biotechnological research of *Phalaenopsis* have been made in the areas of thermo-tolerance, pathogen resistance, flowering control, flower color and virus diagnosis. This strength of research ability and experience should attract international cooperation on technological applications.

Taiwan's *Phalaenopsis* production system has evolved to a commercialized scale by through involvement of many breeders and companies. In addition, it has developed into a complete high tech system, i.e. from product selection, healthy seedling propagation, quarantine systems, and production automation, which form a well-set package in agriculture developments. As such, it is well situated for forming strategic alliances worldwide as a turn key project.

In order to vertically integrate the orchid industry, the Council for Economic Planning and Development, has established and funded the Taiwan Orchid Plantation (TOP) in Tainan County in 2003. TOP was designed and developed by myself together with the management team of A-Turn Biotech. We believe that TOP will be the platform to strengthen the orchid industry in R&D, mass production, exhibition, trading and so forth, which will make Taiwan the largest orchid supplier in the world. In addition, the "TOP" name will provide a strong differentiating brand for Taiwan's *Phalaenopsis*.

Since there are many highly skilled breeders and technology developers on the island, TOP will also be acting as a platform for trading orchid varieties and technologies internationally. Taiwan has the capability to custom-make specific products for every specific market in the shortest possible period because of its rich breeding materials and skillful breeders. This will enhance its position as a global leader of *Phalaenopsis* suppliers.

Technological development is the foundation of the industry; it takes tremendous time and investment to build up, but is absolutely essential. I am glad to see this book published and believe that the book will become the most useful and valuable technological resource for all orchid lovers worldwide.

Dr. Irwin Y.E. Chu
Founder
A-Turn Biotech Co.

Preface

The appreciation of orchid beauty has a long history in both western and eastern cultures. Over many years of development, the orchid has evolved such that it embraces not just the hobbyists' market but a highly commercial market, thanks to advances in techniques such as breeding, micropropagation, industrial cultivation, etc. Today, orchid cut-flowers of *Cymbidium*, *Dendrobium* and *Oncidium*, and potted plants of *Phalaenopsis* are marketed globally. It is envisaged that growing tropical orchids for cut-flower production and potted plants will benefit from the recent advances in the crop science technology. However, for the orchid industry, producing an improved orchid through biotechnology is only the beginning.

Taiwan has been the main driving force of the world's *Phalaenopsis* breeding and plant production. The orchid research program was firstly supported 10 years ago by the Taiwan Sugar Corporation for the first three years, and currently has been one of the National Science-Tech Program for Agriculture Biotechnology (NSTP.AB) for more than six years. The budgets of the NSTP.AB are founded by National Science Council, Council of Agriculture and Academic Sinica, Taiwan. The contributors to the book include researchers from the Institute of Plant and Microbial Biology, Academia Sinica, National Taiwan University, National Tsing Hua University, National Cheng Kung University, National University of Kaohsiung, and National Pingtung University of Science & Technology. We collaborate with the growers of Taiwan Orchid Plantation, a government sponsored entity, in terms of research and training, in order to bring the Taiwan orchid industry to a new level of sophistication and profitability.

This book is the first volume devoted exclusively to orchid biotechnology. It is extremely informative as it addresses many aspects of orchid biotechnology, including modern breeding (Chapters 1 and 2), *in vitro* morphogenesis (Chapter 3), somaclonal variation (Chapter 4), application of orchid mycorrhized fungi (Chapter 5), analysis of orchid genomes (Chapters 6–8) and functional genomics (Chapters 9–12), and genetic transformation (Chapter 13). It will be a valuable guide for readers such as research workers, graduate students, people interested in orchid biology and floriculturists. Its publication will be a milestone sets the foundation for the next level of orchid research.

Wen-Huei Chen and Hong-Hwa Chen
The Editors

Contents

List of Contributors

Ming-Tsair Chan
Agricultural Biotechnology Research Center
Academia Sinica
Taipei
Taiwan

Ching-Chun Chang
Institute of Biotechnology
National Cheng Kung University
Tainan
Taiwan

Doris C. N. Chang
Department of Horticulture
National Taiwan University
Taipei
Taiwan

Wei-Chin Chang
Institute of Plant and Microbial Biology
Academia Sinica
Taipei
Taiwan

Fure-Chyi Chen
Department of Plant Industry and Graduate Institute of Biotechnology
National Pingtung University of Science & Technology
Pingtung
Taiwan

Hong-Hwa Chen
Department of Life Sciences
National Cheng Kung University
Tainan
Taiwan

Wen-Huei Chen
Department of Life Sciences
National University of Kaohsiung
Kaohsiung
Taiwan

Mei-Chu Chung
Institute of Plant and Microbial Biology
Academia Sinica
115, Taipei
Taiwan, ROC

Yu-Yun Hsiao
Department of Life Sciences
National Cheng Kung University
Tainan
Taiwan

Chien-Hao Huang
Department of Botany
National Taiwan University
Taipei
Taiwan

Yen-Yu Kao
Department of Botany
National Taiwan University
Taipei
Taiwan

Institute of Molecular and Cellular Biology
National Taiwan University
Taipei
Taiwan

Chang-Sheng Kuoh
Department of Life Sciences
National Cheng Kung University
Tainan
Taiwan

Hsiao-Ching Lee
Institute of Bioinformatics and Structural Biology and Department of Life Science
National Tsing Hua University
Hsinchu
Taiwan

Yung-I Lee
Institute of Plant and Microbial Biology
Academia Sinica
115, Taipei
Taiwan, ROC

Botany Department
National Museum of Natural Science
Taichung
Taiwan

Yi-Hsueh Li
Department of Botany
National Taiwan University
Taipei
Taiwan

Chih-Chung Lin
Department of Botany
National Taiwan University
Taipei
Taiwan

Chin-Wei Lin
Department of Life Sciences
National Cheng Kung University
Tainan
Taiwan

Hsien-Chia Lin
Institute of Biotechnology
National Cheng Kung University
Tainan
Taiwan

Tsai-Yun Lin
Institute of Bioinformatics and Structural Biology
and Department of Life Science
National Tsing Hua University
Hsinchu
Taiwan

Hsiang-Chia Lu
Department of Plant Pathology and Microbiology
National Taiwan University
Taipei
Taiwan

Zhao-Jun Pan
Department of Life Sciences
National Cheng Kung University
Tainan
Taiwan

Sanjaya
Agricultural Biotechnology Research Center
Academia Sinica
Taipei
Taiwan

Jun Tan
College of Bioinformation
Chongqing University of Post and Telecom
Chongqing
China

Ching-Yan Tang
Department of Life Sciences
National University of Kaohsiung
Kaohsiung
Taiwan

Wen-Chieh Tsai
Department of Biological Science and Technology
Chung Hwa University of Medical Technology
Tainan County
Taiwan

Heng-Long Wang
Department of Life Science
National Kaohsiung University
Kaohsiung
Taiwan

Edward C. Yeung
Department of Biological Sciences
University of Calgary
Calgary
Alberta T2N 1N4
Canada

Hsin-Hung Yeh
Department of Plant Pathology and Microbiology
National Taiwan University
Taipei
Taiwan

Kai-Wun Yeh
Institute of Plant Biology
College of Life Science
National Taiwan University
Taipei
Taiwan

Wun-Hong Zeng
Institute of Biotechnology
National Cheng Kung University
Tainan
Taiwan

Chapter 1

Breeding and Development of New Varieties in *Phalaenopsis*

Ching-Yan Tang[†] *and Wen-Huei Chen*[*,†]

One of the most important strategies to keep Taiwan as the leading producer of *Phalaenopsis* in the world, is breeding and development of new varieties. Pedigree analysis of the 12 most popular white hybrids of *Phalaenopsis* indicated that the tetraploids of *Phal. amabilis* and the hybrid, *Phal.* Doris were used frequently as parents of these hybrids. Besides the standard big flower *Phalaenopsis*, development of novelty varieties, such as the Harlequins and the multi-floral types constitute the new trends in the *Phalaenopsis* breeding programs and markets in the last decade. The somaclonal mutants of *Phal.* Golden Peoker and the wild species, *Phal. equestris* played an important role in the development of these novelty varieties. Breeding for new varieties of *Phalaenopsis* is lengthy and time consuming. New techniques are needed to increase the breeding efficiency of crops having long life cycles. The recent development of molecular markers, such as restricted fragment length polymorphism (RFLP), random amplified polymorphic DNA (RAPD) and DNA amplification fingerprinting (DAF) and their applications in *Phalaenopsis* breeding are discussed and evaluated in this chapter.

1.1 Introduction

The 1980s was the decade that has divided the orchid business of Taiwan into two distinct phases. Before 1980, the cultivation of orchid

[*]Corresponding author.
[†]*Department of Life Sciences, National University of Kaohsiung, Kaohsiung, Taiwan.*

was considered a hobby. Most hobbyists were raising orchids in small scale with simple green-house facilities. With the rapid growth of economy, *Phalaenopsis* orchids became one of the most important commodity in the domestic as well as the international markets. Since 1988, Taiwan Sugar Corporation has started a comprehensive program to modernize *Phalaenopsis* production through intensive research effort, while more modern and well-equipped greenhouses[1] were built to meet the demands of an expanding market which could not be met through the activities of orchid hobbyists. Moreover, *Phalaenopsis* breeding became more professional and was usually well designed as compared with the trial and error approach of the traditional breeding programs.

Another change during this period was the product type in the export markets. Instead of cut flowers, the medium- and large-sized seedlings of selected hybrids became the major items for export. Consequently, *Phalaenopsis* growers in Taiwan had to equip themselves with modern greenhouses as well as facilities for mass production of hybrid seedlings which are normally derived from crosses of two high quality parental varieties. Seeds from the mature capsules are sown by *in vitro* method. Young seedlings developed in the test tube are transplanted into pots which are divided into four groups: small-, medium-, and large-sized seedlings and flowering plants, according to market demands.[2] Progeny test is used to evaluate and select the potential hybrids at different stages of development. To save time, parental plants used for hybridization to produce hybrids are propagated by the mericlone method, while evaluation of the new hybrids is in progress. Therefore, at the final stage of selection of the hybrids, there will be enough parental stocks available for making crosses to produce a large amount of hybrid seedlings for the market.[3]

The *Phalaenopsis* varieties used for breeding are usually divided into two groups — the standard big flower group and the novelty group. The standard big flower group includes the white, pink as well as the varieties with stripes, being derivatives of the white *Phal. amabilis* and the pink *Phal. schilleriana*. The varieties of the novelty group are usually small flowers with special coloration; some have special fragrances, i.e. if *Phal. violacea* is involved in the pedigree. Other parental varieties in this group are *Phal. amboinensis*, *Phal. venosa*, etc. In recent years, the pot varieties which have small but plentiful flowers have become a new market trend. *Phal. equestris* and *Phal. stuartiana* are the common parental varieties of this group.

In general, the breeding programs are designed to improve the size and color of the flowers as well as other characteristics such as, longevity, stalk length, leaf shape, ease of cultivation, disease resistance and the number of viable seeds through the selection of parents for hybridization and so on. Through tremendous efforts in breeding, various types of *Phalaenopsis* varieties with attractive color and graceful appearance (Fig. 1.1) have been developed and the success of the development has made Taiwan one of the most important producers of *Phalaenopsis* in the world.

The growing cycles of *Phalaenopsis* orchids are long, a cycle being 2–3 years. Using the traditional hybridization to transmit useful traits into the commercial varieties is a long process which takes years to achieve. In addition, some species of orchids are cross-incompatible, thereby limiting the work of variety improvement. Hence, new approaches and techniques are needed in order to produce superior *Phalaenopsis* varieties for the fast growing and highly competitive markets. This chapter discusses the recent developments in the breeding work of *Phalaenopsis*.

Fig. 1.1. *Phalaenopsis* varieties showing various attractive colors and graceful appearance.

1.2 Development of *Phalaenopsis* Varieties by Hybridization

1.2.1 *White Phalaenopsis varieties*

The standard big white flower is the most important group of *Phalaenopsis* in the market (Fig. 1.2). Besides the large size, the breeding objectives of the white *Phalaenopsis* include long flower stalk, well-shaped flowers with long life span, etc. Taiwan is located at the northern border of the natural growth habitat of *Phalaenopsis*. A white flowered species, *Phal. amabilis* var. *formosa* was found native in Heng-Chung Peninsula, Taitung County and the Orchid Island off the coast of southern Taiwan.[4] This native species had won several awards in different international orchid conferences as early as in the 1950s for the beauty of their multi-flowers. By using the technique of polyploidization, superior tetraploid varieties with short flower stalk, round-shaped petals and good quality flowers were developed. These varieties were well accepted by the Japanese market.

The modern superior large white hybrids were developed through the hybridization of breeding stocks from different sources, including those from the local *Phalaenopsis* farms and many from foreign countries, such as Japan, the Netherlands and the United States. Based on these materials, large, well-shaped white-flowered *Phalaenopsis*

Fig. 1.2. The appearance the standard big white variety "*Phal*. Taisuco Brinasu."

hybrids with uniform morphology were developed. Through analysis of the pedigree of the 12 most popular white Taisuco *Phalaenopsis* hybrids in 1997/98, it was found that all of them were the offspring of *Phal. amabilis, Phal. rimestadiana, Phal. aphrodite, Phal. schilleriana, Phal. stuartiana and Phal. sanderiana* with the exception of the *Phal.* Taisuco white which was not related to *Phal. stuartiana* and *Phal. sanderiana* (Table 1.1). Among these wild species, *Phal. amabilis, Phal. rimestadiana* and *Phal. aphrodite* were the most important ancestors for the modern white commercial hybrids.[5] The proportion of the genetic constitution contributed by *Phal. amabilis, Phal. rimestadiana* and *Phal. aphrodite* were 40.34%, 38.56%, and 16.41%, respectively. Based on this information, one can note that these hybrids were closely related in their genetic make-up. This narrow genetic background in *Phalaenopsis* white hybrids was difficult to avoid due to the demand for high uniformity of the hybrid seedlings by the market. That means genetic homogeneity of the parental stocks was required in order to produce uniform hybrid seedlings. However, one has to be aware that genetic depression may occur during the process of improvement.

From the same analysis, it was found that these 12 Taisuco hybrids were originated from 17 ancestral hybrids (Table 1.2). However, the

Table 1.1. Genetic Contribution of the Wild Species for the 12 Most Popular Commercial Hybrids[a] of the White Taisuco *Phalaenopsis*

Wild Species Used in the Pedigree	Percentage of Genetic Contribution	Mean of Percentage	C.V.[b]
Phal. amabilis	39.11–42.19	40.34	2.8
Phal. rimestadiana	37.64–39.22	38.56	1.2
Phal. aphrodite	15.36–17.74	16.41	4.3
Phal. schilleriana	2.64–4.49	3.48	19.8
Phal. stuartiana	0–1.17	0.51	70.6
Phal. sanderiana[c]	0–0.78	0.47	51.1

[a]The name of the 12 hybrids are: *Phal.* Taisuco Kochdian, *Phal.* Taisuco Kaaladian, *Phal.* Taisuco Windian, *Phal.* Taisuco Bright, *Phal.* Taisuco Bridian, *Phal.* Taisuco Adian, *Phal.* Taisuco White, *Phal.* Taisuco Brinasu, *Phal.* Taisuco Silver, *Phal.* Taisuco Crane, *Phal.* Taisuco Swan, *Phal.* Taisuco Nasubula.

[b]C.V. = coefficient of variation.

[c]The commercial hybrid, *Phal.* Taisuco White, does not have the genetic contribution of *Phal. stuartiana* and *Phal. sanderiana*.

Table 1.2. Genetic Contribution of the Important Parental Hybrids in the Pedigree of the 12 Most Popular Commercial Hybrids[a] of the White Taisuco *Phalaenopsis*

Name of Parental Hybrid	Genetic Contribution (%)	No. of Commercial Hybrids
Group A		
Phal. Elisabethae	40.23	12
Phal. Gilles Gratiot	30.69	12
Phal. Katherine Siegwart	31.56	12
Group B		
Phal. Doris	50.47	12
Phal. Doreen	17.36	9
Phal. La Canada	12.18	9
Phal. Winged Victory	19.37	12
Group C		
Phal. Grace Palm	23.57	12
Phal. Thomas Tucker	12.26	9
Group D		
Phal. Elinor Shaffer	18.75	10
Phal. Long Life	14.84	8
Phal. Opaline	14.84	8
Phal. Vallehigh	22.27	8
Group E		
Phal. Kochs Schneestern	27.78	9
Phal. Meridian	27.78	9
Phal. Mount Kaala	26.70	11
Phal. Schone Von Unna	14.84	8

[a]The names of the 12 hybrids are the same as in Table 1.1.

average genetic contribution for *Phal.* Doris to the current large white *Phalaenopsis* hybrids was about 50.47%, which was equally important to the direct parental hybrids. Through the genetic flow from the ancestors to the modern white Taisuco hybrids, it is observed that the superior clones of the large white Taisuco *Phalaenopsis* were developed firstly through the improvement of the genetic characters for *Phal.* Doris by chromosome doubling, resulting in a tetraploid with a larger genomic capacity to accumulate more additive alleles. Then it was followed by backcrossing and hybridizing with its relatives to recombine and to accumulate desirable additive alleles for the flower size and other favorable traits. By this selection scheme, more than 30 Taisuco *Phalaenopsis* white hybrids were obtained and they won many awards throughout the world, including eight from the American Orchid Society.

1.2.2 *Harlequin (novelty) varieties*

Development of the Harlequin varieties is a new trend in *Phalaenopsis* breeding which was developed in Taiwan in the last 12 years. The most distinguished characteristic of this new group of *Phalaenopsis* is the appearance of large blotches of coalesced spots with intense color against the light creamy white or other colors. The blotching appears to be unstable. It may vary in size, shape and location from flower to flower. It was also found that temperature may influence the expression of the blotches.[6] With cooler temperatures, the color intensity of the Harlequin spot will increase.

The breeding of the Harlequins began in Taiwan when the famous hybrid *Phal.* Golden Peoker "Brother" (*Phal.* Misty Green × *Phal.* Liu Tuen Shen, Reg. Brothers's Orchid, 1983) was mericloned in the 1990s. The special feature of this variety is its creamy white flower with intense wine-colored spots. From the pedigree analysis, *Phal.* Golden Peoker was developed from 12 wild species through 11 generations.[7] The genetic contribution in generating wine-colored spots is 25, 18.75, 12.5 and 6.25% from *Phal. gigantea*, *Phal. leuddemanniana*, *Phal. amboinensis* and *Phal. faciata*, respectively (Table 1.3). Besides the contribution of nicely spotted flowers in the particular group, these species also have a tendency to add other characters, such as leather-like texture, and flattened and round appearance of the flowers of the Harlequins. In the process of mericloning the *Phal.* Golden Peoker "Brother," somaclonal mutants with different, remarkably, fused spots of Harlequin pattern on the sepals and petals have been found in different orchid farms in Taiwan. Among these mutants, three of them, namely "Ever-spring," "Nan Cho" and "S.J." were most famous and received AOS recognition and awards. The mericlones of these mutants also produced flowers that were dominated with this Harlequin pattern (Fig. 1.3). From the emergence of these clones, intensive breeding for the Harlequins began. They were used to hybridize high quality

Table 1.3. Genetic Contribution of the Wild Species to *Phal.* Golden Peoker

	No. of Times of Each Species Introduced into Each Generation											
Wild Species	**1**	**2**	**3**	**4**	**5**	**6**	**7**	**8**	**9**	**10**	**11**	**% of Genetic Contribution**
P. gigantean		1										25.00
P. luddemanniana			1	1								18.75
P. rimestadiana							4	19	18	9	2	15.04
P. amboinensis			1									12.50
P. amabilis							3	13	12	4		10.16
P. aphrodite							1	9	13	5	2	7.42
P. faciata				1								6.25
P. sumarana						1						1.56
P. schilleriana								2	2			1.27
P. stuartiana							1		1			1.07
P. equestris							1					0.78
P. sanderiana									1			0.20

Data from RHS98 (1998).

Fig. 1.3. Coalescence of red-brownish blotches on the flowers is the characteristic of Harlequin *Phalaenopsis* (*Phal.* Golden Peoker "A87–100").

parental varieties possessing different flower colors and to create novel cultivars of Harlequin flowers.

Phal. Golden Peoker is an excellent parent. From 1992 to 2003, 423 hybrids developed from *Phal.* Golden Peoker were registered in Sander's list of orchid hybrids. Among them, 149 were the Harlequins. In addition to *Phal.* Golden Peoker, there were six important related varieties which were widely used in the breeding programs to create novel cultivars of Harlequins (Table 1.4). From 1994 to 2002, 58 Harlequin flowers won awards from the AOS. Among these varieties, *Phal.* Ever Spring Fairy "Tokai Silky Star" and *Dtps.* Chain Xen Diamond "Celebration" were so commanding that they received 90-points and won the much sought after FCC/AOS. Another variety, "*Dtps.* Ever Spring Prince" received seven AOS awards in 2001 and 2002, including two AMs and five HCCs.

Good progress has been made to breed for Harlequin varieties since the development of *Phal.* Golden Peoker "ES," a somaclonal variant. Due to the fascinating and unpredictable pattern of the Harlequin flowers, there remains a lot of room for the improvement of this group of novelty variety in the future.

Table 1.4. Number of Registered Hybrids Derived from Harlequin Parents in the Breeding Program[a]

Harlequin Variety	Parent	Generation	No. of Hybrids Registered
Phal. Golden Peoker[b]	*Phal.* Misty Green × *Phal.* Liu Tuen-Shen	0	96
Phal. Ever-spring King	*Phal.* Chih Shang's Stripes × *Phal.* Golden Peoker	1	37
Phal. Ever-spring Light	*Phal.* Ever-spring Star × *Phal.* Golden Peoker	1	9
Phal. Ever-spring Prince	*Phal.* Golden Peoker × *Dtps.* Taisuco Beauty	1	8
Phal. Ching Her Prince	*Phal.* Ever-spring King × *Phal.* Golden Peoker	2	6
Phal. Haur Jin Diamond	*Phal.* Golden Peoker × *Phal.* Ching her Buddha	1	5
Phal. Ho's Fantastic Splash	*Phal.* Ever-spring King × *Phal.* Ho's French Fantasia	2	5

[a]Data from Wildcatt Orchids Database (2003) and RHS98 (1998).

[b]No. of crosses including all clones of *Phal.* Golden Peoker which have "Brother", "Ever-spring", "Nan-Cho", "S.J." and "BL", etc.

1.2.3 *Potted varieties*

Before the 1980s, cut-flowers dominated the *Phalaenopsis* market. However, demands for potted *Phalaenopsis* varieties have increased tremendously in the last decade. Breeding for the potted-plant market is different from breeding for the cut-flower market. While the cut-flower market emphasized floral traits, for the potted-plant market, vegetative traits are equally important. These traits include small plants with considerable number of flowers, shortened and multiple branching of the inflorescence, easy growing and flowering, etc. At the beginning, potted varieties available for the market were usually smaller version of the standard *Phalaenopsis*. Selections were made for more compact growth and flowering. As the demand of potted varieties increased, special effort was made to develop hybrids for this sector of the market. For this purpose, *Phal. equestris* is being used as the most important parent in producing hybrids for potted varieties.[8] *Phal. equestris* is a native species in the Philippines and is one of the two indigenous *Phalaenopsis* species in Taiwan. Bearing either white or pink flowers are two common forms of *Phal. equestris*. Sequential flowering having flowers all around the inflorescence and short flower stalks are special characteristics of this species. In addition to *Phal. equestris, Phal. stuartiana* and *Phal. schilleraiana* are also important species used for breeding of potted varieties having heavily branching flowers. The first important hybrid in this line of breeding is *Phal.* Cassandra (*Phal. equestris* × *Phal. stuartiana*). Though it was made by Seden and was registered by Veitch in 1899,[9] it was not heavily used in *Phalaenopsis* breeding until the 1960s. Numerous hybrids with multi-branching which is expected as a major characteristic of multi-floral varieties, were developed by using *Phal.* Cassandra as one of the parents in their pedigree. More than 150 first generation hybrids were registered using Cassandra as one of the parents. Recently, Cassandra hybrid was remade by crossing the tetraploid form of *Phal. equestris* "Riverbend" and *Phal. stuartiana*. The resulting hybrid was a triploid hybrid which was sterile. The tetraploid form of the other parental varieties is needed in order to make use of the advantages of tetraploid breeding. Breeding for potted *Phalaenopsis* varieties is a new trend in the markets. One can expect to see more and better varieties in the near future.

1.3 Breeding Behavior and Inheritance

1.3.1 *Inheritance of floral color*

Taiwan is one of the native habitats of *Phal. amabilis* and *Phal. equestris* which are used extensively in the development of *Phalaenopsis* hybrids.[10] *Phal. equestris* which appeared naturally with pink or white flowers, produces branched inflorescences with a short juvenile period, and is naturally dwarf. It is an important parent for breeding the miniature type of plants which produce a large number of small flowers and are much easier to pack and to transport. It was also used to produce hybrids that had white petals and sepals with a red lip (semi-alba). Plants of *Phal. equestris* are highly variable in terms of morphology as well as in floral color. It can be divided into the following forms[11]:

(1) *Phal. equestris* var. *alba* — a pure white form; no yellow pigments on the callus.
(2) *Phal. equestris* var. *aurea* — white flowers with solid yellow lip.
(3) *Phal. equestris* var. *leucotante* — flowers with white lips and yellow callus.
(4) *Phal. equestris* var. *rosea* — flowers with even red petals and sepals; color of the mid-lobe of the lip varies from deep red to light red.
(5) *Phal. equestris* var. *leucaspis* — small flowers with white edges on pink petals and sepals; mid-lobe of the lip is purple or orange in color with white or yellow callus.

It was reported that several independent genes control the colors of the lip of *Phal. equestris* through the expression of both anthocyanins and carotenoids.[12] By crossing between the white and red forms of *Phal. equestris*, Fu *et al.*[11] reported that the pink floral color was controlled by a single dominant gene. This gene acts on the coloration of petals, sepals, the mid-lobe and the apex area of the side-lobe of the lip. Also, it is expressed as a brownish color through a pleiotrophic effect on the coloration of the floral stalk and the spot of the callus. Compared with the petals and sepals, the color inheritance of the lip is more complicated. Two dominant duplicated genes which are independent of the above mentioned red gene, control the yellow color of the base of the mid-lobe and side-lobe as well as the callus of the lip.

Since *Phal. equestris* is frequently used to cross with other hybrids, it has been found that the expression of these color-genes varies according to the different genetic backgrounds. For example, when a pink-flowered *Phal. equestris* was used to cross with various commercial hybrids of pink, yellow with magenta spots or semi-alba floral colors, the colors of the flower of the progenies were pink, orange with pink blush, lavender or white with pink splash, respectively. Retention of the pink color in the flowers of the progenies from crosses with different genetic backgrounds suggested that the inheritance of pink floral colors of *Phal. equestris* might be controlled by a dominant gene[13] which was the same red gene as in the previous study. However, when the same *Phal. equestris* was used to cross with commercial hybrids of the white, orange or yellow varieties, flowers with pink lips and various degrees of pink blush were observed in the progenies. These results suggested the presence of two complementary genes, C and R which controlled the pink color in the flowers of *Phal. equestris*, similar to those in *Cattleya*.[14]

1.3.2 *Influence of Phalaenopsis equestris parents on fertility*

To study the relationship between fertility and pollen or pod parents used in the crosses, four varieties (including two white and two pink forms) of *Phal. equestris* were used to cross with the commercial hybrids. Each form was used as either pollen or pod parents and *vice versa* for the commercial hybrids. A total of 147 crosses were made and the fertility of each cross was determined by measuring the viable seeds produced from each cross. The results showed that 50–57% of the crosses (Table 1.5) produced viable seeds if the white or pink forms of *Phal. equestris* were used as pollen parents to cross with the commercial hybrids. However, no viable seed was produced if *Phal. equestris* was used as pod parents, regardless of the floral color. The varieties of *Phal. equestris* used in this study were diploids while the majority of the other parents were tetraploid hybrids. That means failure of seed production was found when the tetraploid plants were used as pollen parents to cross with the diploid varieties. Therefore, in order to enhance the breeding efficiency, a breeder has to use the diploid varieties as pollen parents if the counterparts are tetraploid plants.

Table 1.5. Effect of *Phalaenopsis equestris* Parents on the Fertility

Floral Color of *Phal. equestris*	*Phal. equestris* Used as Pod or Pollen Parents	Total No. of Crosses	Fertility > 0[a]	
			No. of Crosses	% of Crosses
White	Pollen parent	36	18	50
White	Pod parent	10	0	0
Pink	Pollen parent	91	52	57
Pink	Pod parent	10	0	0

[a]The crosses with viable seeds were considered as fertility > 0.

1.4 Application of Molecular Markers in *Phalaenopsis* Breeding

1.4.1 *Screening for red floral gene by RAPD markers*

A single dominant gene was found to control the red floral color (as against white color) in *Phal. equestris*.[11] Due to the long life cycle of *Phalaenopsis* orchids, it is a time consuming procedure to identify the progenies carrying this gene after hybridization. Therefore, a rapid technique is needed for early detection of the presence of this gene in order to increase the efficiency of the breeding procedure. By using a stepwise screening method, Chen *et al.*[15] identified a RAPD (random amplified polymorphic DNA) marker linked to the red floral gene in *Phal. equestris*. In this experiment, the leaf tissue of plants from the white and red parents, F1 and F2 progenies were subjected to RAPD analysis. In the first step, 920 primers were screened by RAPD analysis using the leaf-tissue from the white and red parents and a single F1 plant. One hundred and fifty (16.3%) of them were found to produce distinct DNA polymorphic bands in the red floral parent and F1 progeny. These were absent in the white floral parent (Table 1.6). In the second step, three F1 plants were used for the detection of polymorphic and homozygous bands that could distinguish the red floral parent and F1 progenies from the white floral parent. Homozygous trait could be confirmed if polymorphic bands were present in all three F1 plants. Among the 150 primers selected from the first step, 34 showed distinct polymorphic and homozygous DNA bands that could distinguish the red

Table 1.6. Numbers and Probabilities of Polymorphic Primers Detected by RAPD Analysis and PCR Reactions Required in the F1 and F2 Progenies of *Phalaenopsis equestris* from 920 Primers

	F1 Plants		F2 Plants
	First Screening	Second Screening	Third Screening
No. of polymorphic primers	150	34	1
Probability of polymorphism (%)	16.3	3.7	0.1
No. of PCR reactions	2,760	750	3,604

flowered parent and the three F1 progenies from the white flowered parent. The third screening was made by using 34 primers for 106 individual plants (84 from red, 22 from white) randomly selected from the F2 population. The results showed that the primer "OPQ-10" (5′-TGT-GCC-CGA-A-3′) generated a 380 bp DNA band (OPQ 10–380) that was linked to the red floral gene. Chi-square analysis indicated that the OPQ 10–380 marker and the red flower gene were two closely linked genes with a distance of 30.8 centiMorgan (cM) apart. With the use of this OPQ 10–380 marker, one can identify the presence of the red-flower gene in the progenies of hybridization at any stage of the plant development. In association with the *in situ* hybridization technique, the molecular marker may potentially be used for the identification and gene mapping of the chromosome where the red flower gene is located.

1.4.2 *Investigation of the parental and phylogenetic relationship by RFLP markers in chloroplast DNA*

Traditionally, morphological characteristics and cytological analysis were used for the classification of plant species as well as for the phylogenetic study. Recently, due to the fast development of biotechnology at the molecular level, molecular markers using restriction fragment length polymorphism (RELP) or random amplified polymorphic DNA (RAPD) were commonly used for various areas of plant sciences.[16,17] Because of the specificity, consistency and precision of the performance of these molecular markers, these techniques became widely used to study the phylogenetic relationship of plant species or the parental relationship in the plant breeding programs.

In one of the studies, RFLP was used to analyze the mode of inheritance of chloroplasts in both interspecific hybrids of *Phalaenopsis* (*Phal. amabilis* × *Phal. amboinensis*; *Phal. mannii* × *Phal. stuartiana*) and intergeneric hybrids of *Phalaenopsis equestris* and *Doritis pulcherrima.*[18] Chloroplast DNA digested with *Dra* I followed by hybridization with an *rbc*L probe revealed that *Phal. amabilis, Phal. aphrodite* and *Phal. stuartiana* had the same size 2.0-kb fragment while the *Phal. mannii* and *Phal. amboinensis* had a 2.3-kb fragment. The size of the fragment in *Doritis pulcherrima* was 3.5-kb. In the analysis of the interspecific reciprocal crosses between two *Phalenopsis* species or the intergeneric reciprocal crosses between *Phal. equestris* and *Doritis pulcherrima*, similar results were found, i.e. the sizes of the fragments shown in the F1 progenies were the same as that in the maternal parents (Table 1.7). Therefore, maternal inheritance of the cpDNA as revealed by the RFLP markers was clearly demonstrated in the reciprocal crosses between the interspecific hybrids and intergeneric hybrids. These results suggested that cpDNA can be used as a marker for the identification of the parentage and for phylogenetic studies of taxonomy.

1.4.3 *Use of RAPD markers for phylogenetic study and variety identification*

By using the morphological characteristics of the petal and sepal, Sweet[19] classified *Phalaenopsis* orchids into 45 species and 9 sections. All the species of *Phalaenopsis* have the same chromosome number ($2n = 38$) with chromosome sizes ranging from 1.5 to 3.5 μm.[20] They can be divided into large, medium and small chromosome groups according to their chromosome size.[21] By using flow-cytometry, Lin *et al.*[22]

Table 1.7. Polymorphism as Shown by the RFLP Markers in the Parental Lines and the F1 Progenies of the Interspecific Reciprocal Crosses of *Phalaenopsis*

	Parents		F1 Progenies	
Fragment Size (kb)	**A[a]**	**B**	**A × B**	**B × A**
2.3	+[b]	−	+	−
2.0	−	+	−	+

[a]A and B represent *Phal. amboinensis* and *Phal. amabilis*, respectively.
[b]+ and − represent presence or absence of polymorphism as shown by RFLP markers.

studied the nuclear DNA content of 18 species of *Phalaenopsis*. The quantities of the nuclear DNA content ranged from 2.74–16.61 pg/2c. They were classified into eight groups according to the nuclear DNA content. This information is useful in terms of orchid classification as well as the phylogenetic relationship among species. In addition to the various approaches mentioned, RAPD analysis was also used for these purposes. Fu *et al.*[23] studied the relationship of 16 wild species of *Phalaenopsis* using RAPD markers. They found that the similarity coefficient and the relative order were stabilized when 20 primers were used to generate 381 DNA bands for analysis. By using the results of this analysis, 16 wild species of *Phalaenopsis* could be classified into five groups (Table 1.8) according to the similarity

Table 1.8. Comparison of the Classifications of 16 Wild Species of *Phalaenopsis* According to the Dendrogram Generated by RAPD,[23] Morphological Characteristics[19] and Chromosome Sizes[20]

Species	Group According to RAPD Data[a]	Section According to Sweet, 1980	Group According to Chromosome Size
Phal. micholitzii	E	Amboinensis	NA[b]
Phal. intermedia	E	Phalaenopsis	NA
Phal. mannii	D	Polychilos	Large
Phal. lueddemanniana	D	Zebrinae	Medium
Phal. mariae	D	Zebrinae	Small
Phal. pulchra	D	Zebrinae	NA
Phal. sumatrana	C	Zebrinae	NA
Phal. venosa	C	Amboinensis	Medium
Phal. violacea	C	Zebrinae	Large
Phal. gigantea	B	Amboinensis	Medium
Phal. amboinensis	B	Amboinensis	Medium
Phal. schilleriana	A	Phalaenopsis	Small
Phal. stuartiana	A	Phalaenopsis	Small
Phal. amabilis	A	Phalaenopsis	Small
Phal. aphrodite	A	Phalaenopsis	Small
Phal. equestris	A	Stauroglottis	Small

[a]Grouping according to Fu *et al.*[23]
[b]NA = not available.

coefficient and the relative order as shown by the dendrogram. The authors claimed that 11 out of 16 species studied were matched between the grouping methods based on morphological characteristics and use of molecular markers. Furthermore, *Phal. amabilis* and *Phal. equestris* were the most closely related species according to the RAPD data, but they were classified into two far-related sections based on morphology. Similarly, *Phal. mannii* and *Phal. lueddemanniana* were considered to be closely related according to the RAPD data which was different from the traditional taxonomic classification. If the cytogenetic evidence comes into the picture, one can find that the chromosome size of *Phal. amabilis* and *Phal. equestris* falls into to the small group, while those of the *Phal. mannii* and *Phal. lueddemanniana* falls into to the large group. It is more reasonable to put the varieties having similar chromosome size into the same group as shown by the RAPD data, instead of into different groups as in the traditional classification based morphological characteristics. In addition, based on comparison between the dendrogram generated by the RAPD analysis and chromosome sizes as shown by the study of the karyotype,[20,21,24] it is noted that the tendency is for the *Phalaenopsis* chromosome size to probably evolve from large to small, and its origin seems to be polyphyletic.

Cross-incompatibility is one of the problems needed to be solved in the *Phalaenopsis* breeding programs. Compatibility is usually correlated to the closeness of their phylogenetic relationship which is, on the other hand, related to the status of the chromosome (i.e. chromosome number and size) and the homology of the nuclear DNA. On the other hand, RAPD analysis as shown by the previous study provides a rapid method to understand the phylogenetic relationship of different species. If this relationship is correlated with compatibility among species, it becomes an useful reference for the choice of parents in the work of hybridization by the breeders.

The morphological characteristics, and cytological and isozyme analysis were generally used in the identification of new species and cultivars. However, these methods are limited by the environmental effects and the diagnostic resolution. Recently, DNA amplification fingerprinting (DAF) has been shown to be an effective method in detecting polymorphism and thus is a powerful tool for species or cultivar identification.[25] In a study, 20 random primers were used to

Table 1.9. DAF Patterns Generated by 20 Random Primers which could Distinguish among 5 Genera,[a] 5 Species in *Phalaenopsis* and 5 Clones in *Phal. equestris*

Primer	Genera	Species	Clones	Primer	Genera	Species	Clones
OPF-1	–[b]	+	+	OPF-11	–	–	–
OPF-2	+	+	–	OPF-12	+	–	–
OPF-3	–	–	–	OPF-13	–	–	–
OPF-4	–	+	+	OPF-14	–	–	–
OPF-5	+	–	–	OPF-15	+	+	–
OPF-6	+	–	–	OPF-16	–	+	–
OPF-7	+	–	–	OPF-17	–	+	–
OPF-8	+	–	–	OPF-18	–	+	–
OPF-9	+	–	–	OPF-19	–	–	–
OPF-10	+	+	–	OPF-20	–	–	+

[a]5 genera were: *Phalaenopsis, Doritis, Cattleya, Dendrobium, Cymbidium*; the 5 species of *Phalaenopsis* were: *Phal. amabilis, Phal. amboinensis, Phal. mannii, Phal. violacea, Phal. equestris*; 5 clones of *Phal. equestris*.

[b]+ and – represent presence or absence of polymorphism as shown by RAPD markers.

analyze the DAF patterns among five genera, five species in the genus of *Phalaenopsis* and five clones in a species, *Phalaenopsis equestris*. Polymorphism was observed among them when a suitable primer was used in the PCR reaction. In this study, it was shown that 9, 8 and 3 primers produced considerable polymorphism which could distinguish among five genera, five species and five clones, respectively (Table 1.9). Distinguishable bands of DAF patterns among the clones with similar genetic background were obtained when a suitable primer was used. Therefore, DAF is a powerful and useful tool to generate a group of molecular markers which represent the identity of a new variety. It is one of the means by which one can use to protect the patent rights of the new varieties in *Phalaenopsis* as well as in other species.

1.5 Conclusion and Prospective

The standard white *Phalaenopsis* is a successful "research and development" product from both the horticultural and industrial points of view. Because of the development of the superior white *Phalaenopsis* varieties, not only did it lead to the opening of an international market for the *Phalaenopsis* business, but it also stimulated the modernization of the production facilities, technique and management for the orchid industry in Taiwan in the last two decades. In this review, one can find that the success of the standard white *Phalaenopsis* varieties was based on the discovery of the tetraploid from *Phal. amabilis* and the hybrid *Phal.* Doris. From the analysis of the 12 white TAISUCO varieties, these two tetrapoloids were involved in their pedigree one way or the other. This means the development of white *Phalaenopsis* is no longer at the diploid level; it is a kind of tetraploid breeding. Although there are only a few of tetraploid parental stocks, yet new and superior varieties of standard white *Phalaenopsis* varieties have been developed year after year. This indicates that the genetic heterogeneity of the tetraploid parents is broad enough to maintain the genetic variability for continuous selection. However, one cannot overlook the potential problem of genetic depression due to the narrow genetic background in these stocks. Exploration and development of new tetraploid breeding stocks with diverse genetic background is an urgent need in order to develop better white *Phalaenopsis* varieties as well as other types of moth orchid.

Development of the novelty varieties, including Harlequin and multifloral *Phalaenopsis* is a new trend in the orchid business since the last decade. It opens the door to the exploitation of the use of the genetic diversity in various wild species of *Phalaenopsis* to create new types of varieties besides the standard moth orchids. This approach in *Phalaenopsis* breeding including various kinds of interspecific and intergeneric crosses will form more diverse and unusual types of *Phalaenopsis* that may be important in the future market. Because of the creation of novelty varieties, demands for *Phalaenopsis* orchids should continue to grow in the future.

Phalaenopsis breeding is a lengthy and time consuming process due to the long life cycle. DNA markers associated with useful genes such as the red floral gene of *Phal. equestris* as reviewed in this chapter, will increase the breeding efficiency through identification of the desired offspring at the seedling stage. Use of the RFLP and RAPD markers to study the phylogenetic relationship of wild species or the parentage of breeding stocks will provide good information on the genetic relationship among different species and cultivars. This kind of information is helpful for breeders to choose the parents for hybridization with more precision. The technique of DNA amplification fingerprinting (DAF) is useful to identify different varieties developed in a breeding program. With this method, a breeder can protect the patent rights of the *Phalaenopsis* hybrids produced.

Looking into the future, there will be unlimited opportunities for the expansion of *Phalaenopsis* orchid in the international markets. However, in order to maintain the competitiveness of Taiwan in the *Phalaenopsis* business, development of new and superior varieties is the key to success. Besides the traditional hybridization technique, effort on the exploration of new sources of genetic diversity as well as the development in the biotechnology of *Phalaenopsis* orchids to increase the breeding efficiency and accelerate the development of novelty varieties should be emphasized, so as to maintain the leading role of Taiwan in the international orchid business.

References

1. Huang TJ, Pan CL, Jean MH, Chen CF. (1996) Design and development of a greenhouse automatic environmental control system. Report of the Taiwan Sugar Research Institute **153:**53–74.
2. Chen WH, Wang YT. (1996) *Phalaenopsis* orchid culture. *Taiwan Sugar* **43:**11–16.
3. Chen WH, Chyou MS, Wu CC, *et al.* (1998) Breeding *Phalaenopsis*. Experimental Report of the Taiwan Sugar Research Institute **1997/1998:**94–102.
4. Lin TP. (1977) *Native Orchids in Taiwan*. Vol. 2. Chong Tao Company, Chiayi, Taiwan.
5. Chen WH, Chen YH, Chyou MS, *et al.* (1999) Development of white Taisuco *Phalaenopsis*. In: Clark J, Elliott WM, Tingley G, Biro J (eds.), *Proceedings of the 16th World Orchid Conference*. Vancouver, Canada, pp. 272–278.
6. Fighetti C. (2004) Passing the torch. *Phalaenopsis-J Int Phalaenopsis All*, Winter 2004:20–31.
7. Chen WH, Chen TC, Wu WL. (2004) The influence of *Phalaenopsis* Golden Peoker "Brother" on Harlequins. *Phalaenopsis-J Int Phalaenopsis All*, Summer 2004:14–16.
8. Harper T. (1991). Mutliflora *Phalaenopsis*: The contribution of *Phal. equestris* in breeding multifloras. *Amer Orch Soci Bull* **60:**106–114.

9. Griesbach RJ. (2002) Development of *Phalaenopsis* orchids for the mass-market. In: Janick J, Whipkey A (eds.), *Trends in New Crops and New Uses*. ASHS Press, Alexandra, VA.
10. Stubbings J. (2006) Development of white with colored lip *Phalaenopsis*. In: Hwang JH (ed.), *Proceedings of Taiwan International Orchid Symposium*, Taiwan Orchid Growers Association, Tainan, Taiwan, pp. 38–51.
11. Fu YM, Chen WH, Tsai WT *et al.* (1996) Studies on floral color heredity of *Phalaenopsis equestris*. Report of the Taiwan Sugar Research Institute **152:**35–49. (In Chinese with English abstract).
12. Christense EA. (2001) *Phalaenopsis*: *A Monograph*. Timber Press, Inc., Portland, Oregon.
13. Chen WH, Tsai WT, Chyou MS, *et al.* (2000) The breeding behavior of *Phalaenopsis equestris* (Schauer) Rchb.f. *Taiwan Sugar* **47(1):**11–14.
14. Lenz LW, Wimber DE. (1959) Hybridization and inheritance in orchids. In: Wither CL (ed.), *The Orchids: A Scientific Survey*. Ronald Press, New York, pp. 261–314.
15. Chen WH, Fu YM, Lin YS, Chen YH. (2001) Identification of RAPD markers linked to the red floral gene in *Phalaenopsis equestris* by a stepwise screening method. *Taiwan Sugar* **48(4):**23–29.
16. Williams JGK, Kubelik AR, Livake KJ, *et al.* (1990) DNA polymorphisms amplified by arbitrary primers are useful as genetic markers. *Nucl Acids Res* **18:**6532–6535.
17. Paran I, Kesseli R, Michelmore R. (1991) Identification of restriction fragment length polymorphism and random amplification polymorphic DNA markers linked to downy mildew resistance genes in lettuce using near-isogenic line. *Genome* **34:**1021–1027.
18. Chang SB, Chen WH, Chen HH, *et al.* (2000) RFLP and inheritance patterns of chloroplast DNA in intergeneric hybrids of *Phalaenopsis* and *Doritis*. *Bot Bull Acad Sin* **41:**219–223.
19. Sweet HR. (1980) The Genus *Phalaenopsis*. *The Orchid Digest*, Inc. USA.
20. Arends JC. (1970) Cytological observation on genome homology in eight interspecific hybrids of *Phalaenopsis*. *Genetica* **41:**88–100.
21. Shindo K, Kamemoto H. (1963) Karyotype analysis of some species of *Phalaenopsis*. *Cytologia* **28:**390–398.
22. Lin S, Lee HC, Chen WH, *et al.* (2001) Nuclear DNA contents of *Phalaenopsis* species and *Doritis pulcherrima*. *J Amer Soc Hort Sci* **126(2):**195–199.
23. Fu YM, Chen WH, Tsai WT, *et al.* (1997) Phylogentic studies of taxonomy and evolution among wild species of *Phalaenopsis* by random amplified polymorphic DNA markers. Report of the Taiwan Sugar Research Institute **157:**27–42. (In Chinese with English abstract).

24. Sagawa Y. (1962) Cytological studies of the genus *Phalaenopsis*. *Amer Orchid Bul* **31**:459–465.
25. Chen WH, Fu YM, Hsieh RM, *et al.* (1995) Application of DNA amplification fingerprinting in the breeding of *Phalaenopsis* orchid. In: Terzi M, *et al.* (eds.), *Current Issues in Plant Molecular and Cellular Biology*. Kluwer Academic, Netherlands, pp. 341–346.

Chapter 2

Embryo Development of Orchids

Yung-I Lee[*,†,‡], *Edward C Yeung*[§] *and Mei-Chu Chung*[†]

The pattern of orchid embryo development is unique among flowering plants. The minute size of the embryos, lack of cotyledon, absence of an endosperm, the varied suspensor morphology, and the simple seed coat structure are some of the unique features of orchid seeds. This chapter summarizes the recent observations on the structural and physiological aspects of orchid embryo development using a few case histories, i.e. *Cypripedium formosanum*, *Calanthe tricarinata* and *Phalaenopsis amabilis* var. *formosa*. The unique features of orchid embryo development such as: 1) the lack of a clear histodifferentiation pattern; 2) presence of a cuticle over the embryo proper; 3) absence of a cuticle in the suspensor cell wall; 4) a suspensor having a transfer cell morphology; and 5) accumulation of high levels of ABA in mature seeds of some terrestrial species are discussed. A comprehensive understanding of the structural and physiological changes during the orchid embryo development will facilitate successful micropropagation of orchids.

2.1 Introduction

Compared to a majority of flowering plants, the process leading to and the pattern of seed development in orchids are unique. In orchids, ovules are not present or poorly developed at the time of anthesis.

[*]Corresponding author.
[†]*Institute of Plant and Microbial Biology, Academia Sinica, 115, Taipei, Taiwan, ROC.*
[‡]*Botany Department, National Museum of Natural Science, Taichung, Taiwan.*
[§]*Department of Biological Sciences, University of Calgary, Calgary, Alberta T2N 1N4, Canada.*

A successful pollination event triggers ovule development within an ovary. Numerous pollen tubes penetrate into the ovary chamber. Once the ovules mature, double fertilization results in the formation of a zygote and a polar chalazal complex within the endosperm cavity of each fertilized ovule.[1,2] The triggering of ovule development after a successful pollination event may ensure the survival of the parent plant since little energy is channeled to the unpollinated ovaries. The polar-chalazal complex fails to develop into an endosperm. Hence, an endosperm is absent from mature orchid seeds. Numerous seeds are produced within a single capsule (Fig. 2.1A). The seeds, which are very small, contain a simple embryo with no distinct tissue differentiation (Fig. 2.1B).[1,3–5] Seed morphology varies within the orchid family.[3] The integuments develop into a thin seed coat with varied surface features.[3,4] Under natural conditions, germination of the mature orchid seeds, especially those from the terrestrial orchids, is dependant upon the association of a mycorrhizal fungus. Since Knudson's discovery,[6]

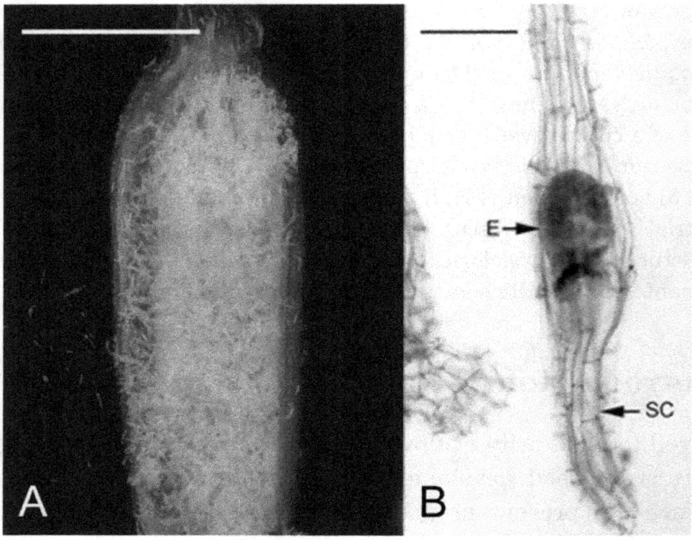

Fig. 2.1. The mature capsule and seed of *Calanthe tricarinata*.

(**A**) Light micrograph showing a mature capsule of *Calanthe tricarinata*. There are numerous tiny seeds within a capsule. Scale bar = 15 mm.

(**B**) Light micrograph showing a mature seed of *Calanthe tricarinata*. SC = seed coat; E = embryo. Scale bar = 150 μm.

orchid seeds can also germinate successfully in a culture medium in the absence of a mycorrhizal fungus. This method is known as asymbiotic germination, and is a useful propagation technique for most orchids. Substantive information on orchid embryo development can be found in a number of reviews.[3–5,7–9] In this chapter, we summarize recent observations on the structural and physiological aspects of orchid embryo development using several case histories as examples. Some of the unique characteristics of embryo development are discussed.

2.2 Embryo Development

2.2.1 *Case histories of embryo development*

2.2.1.1 *Cypripedium formosanum*

Cypripedium formosanum is a terrestrial, native orchid species in Taiwan. At 60 days after pollination, fertilization has just taken place, and an elongated zygote (Fig. 2.2A) can readily be detected within the embryo sac in orchids grown in Mei-Fong farm (2100 m above sea level). Compared with other orchids (Table 2.1), the size of the embryo sac of *Cypripedium formosanum* is large (approximate 62×154 μm). In this species, the endosperm fails to develop. The polar-chalazal complex degenerates early in the endosperm cavity. The growing embryo gradually expands and fills the endosperm cavity. Judging from the location of the mitotic apparatus, the first division of the zygote is asymmetrical (Fig. 2.2A), giving rise to two daughter cells of different sizes and fates (Figs. 2.2B). The smaller terminal cell forms the embryo proper and the larger basal cell gives rise to a short-lived embryonic organ known as the suspensor. The suspensor of this species consists of a highly vacuolated, single cell. The two-celled embryo divides further, resulting in the formation of a four-celled embryo proper at 75 days after pollination (Fig. 2.2C). During globular embryo formation (90–105 days after pollination), cell divisions occur in the outermost as well as the inner layers of the embryo proper, resulting in an increase of embryo volume (Fig. 2.2D and E). Throughout proper embryo development, no further change occurs in the suspensor as cell division or swelling of the suspensor cannot be observed. The suspensor finally degenerates at the late globular stage (Fig. 2.2F). In contrast to the highly vacuolated nature of the suspensor

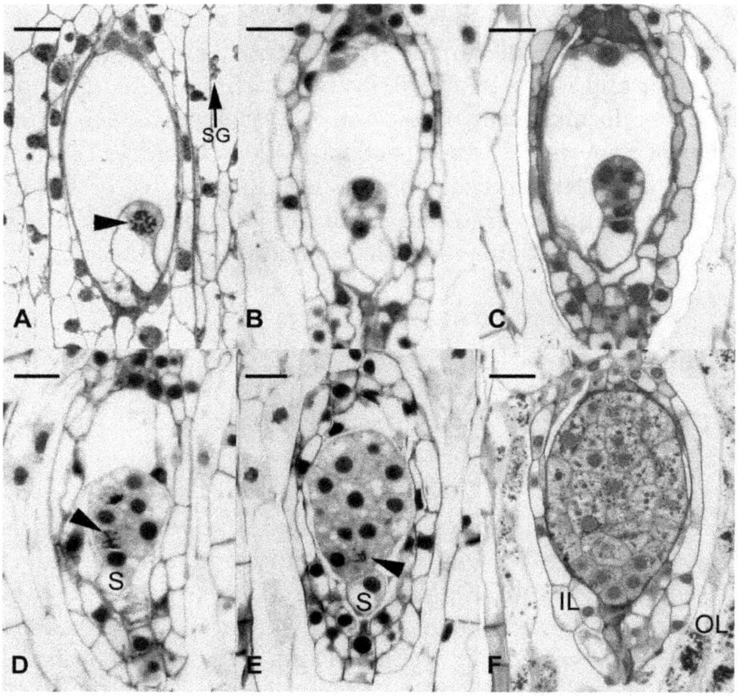

Fig. 2.2. The embryo development of *Cypripedium formosanum*.

(**A**) Light micrograph showing a longitudinal section of a zygote at 60 days after pollination. Before the first transverse division, the condensed chromosomes at metaphase (*arrowhead*) are located towards the chalazal end, while a large vacuole is located at the micropylar end of the cell. SG, starch grains. Scale bar = 30 μm.
(**B**) Light micrograph showing a two-celled embryo. The first cell division of the zygote is unequal, resulting in the formation of a smaller terminal cell and a larger basal cell. Scale bar = 30 μm.
(**C**) Light micrograph showing a five-celled embryo at 75 days after pollination. Scale bar = 30 μm.
(**D**) Light micrograph showing a longitudinal section through an early globular embryo with a single-celled suspensor (S) at 90 days after pollination. The occurrence of periclinal division (*arrowhead*) within the embryo proper results in the formation of the inner tier of cells. Scale bar = 25 μm.
(**E**) At 105 days after pollination, the globular embryo continues to develop by cell division (arrowhead) in the inner layer cell of the embryo proper. The suspensor cell (S) is highly vacuolated. Scale bar = 25 μm.
(**F**) As the seed approaches maturity (150 days after pollination), large vacuoles begin to be replaced by small ones. Small protein bodies begin to appear within the cells of the embryo proper. The cells of both inner (IL) and outer (OL) layers of the seed coat become dehydrated and are gradually compressed into a thin layer. Scale bar = 45 μm.

Table 2.1. The Characteristics of the Embryos of Three Orchid Species in the Case Histories

	Cypripedium formosanum	***Calanthe tricarinata***	***Phalaenopsis amabilis var. formosa***
Habitat	Terrestrial	Terrestrial	Epiphytic
Embryo sac size (X/Y-axis, μm)[a]	62 ± 7.3/154 ± 10.2	31 ± 3.2/54 ± 4.6	18 ± 1.8/30 ± 2.4
Embryo size (X/Y-axis, μm)[b]	125 ± 11.4/ 228 ± 16.2	132 ± 8.5/ 184 ± 14.7	92 ± 10.1/ 250 ± 22.2
Suspensor	1-celled, not enlarged	1-celled, enlarged	8-celled, tubular
A gradient of cell size within embryo	No	No	Yes

[a, b] X, Y ± standard deviation. At least 20 individual embryo sacs and mature embryos are measured.

cell, the cytoplasm of the embryo proper cell is densely stained with amido black 10B — a protein stain. As the seed approaches maturity (150 days after pollination), the vacuoles begin to break down, and some small protein bodies begin to appear within the cells of the embryo proper (Fig. 2.2F).

In this species, the seed coat is derived from the outer integument, which consists of two cell layers. The outer integumentary cells are highly vacuolated and contain chloroplasts with starch granules (Fig. 2.2A). A secondary wall is added to the seed coat cells as the seeds mature. The inner integument is composed of two layers of compact and vacuolated parenchyma cells (Fig. 2.2F). As the seed approaches maturity, the seed coat layers begin to desiccate and are compressed into a thin layer. The dehydrated inner integument is known as "carapace," which is found in some terrestrial orchid species, such as *Cephalanthera*, *Cypripedium* and *Epipactis*. The carapace tightly envelops the embryo at seed maturity. It has been suggested that the carapace can play a role in seed dormancy.[10,11] It is worthwhile to note that both the seed coat and carapace stained blue green with the TBO stain and react positively to the Nile red stain. The accumulation of cuticular material may increase the hydrophobic characteristic of the seed coat that impedes the uptake of water and nutrients during seed germination.[12,13]

During the course of embryo development, the ABA level is low at the proembryo stage, then increases quickly through the late globular stages (120–150 days after pollination), and maintains a maximal level (approximate 13.8 ng · mg^{-1} of fresh weight) at seed maturity.[14]

2.2.1.2 *Calanthe tricarinata*

In *Calanthe tricarinata*, the elongated zygote appears as a highly polarized cell at 70 days after pollination (Fig. 2.3A). Under the light microscope, the nucleus and most cytoplasm of the zygote are located toward the chalazal end, and a large vacuole is found near the micropylar end of the cell. The nucleolus is distinct at this stage. The endosperm of this species degenerates in the early stages of embryo development. After fertilization, the polar nuclei and one of the male nuclei forms the endosperm complex, and these nuclei eventually disintegrate (Fig. 2.3A and B). The first zygote division is unequal, producing a larger basal cell and a smaller terminal cell (Fig. 2.3B). Derivatives of the basal cell give rise to the suspensor, and the terminal cell gives rise to the embryo proper. Later, the terminal cell undergoes further anticlinal and periclinal divisions, resulting in the formation of a globular-shaped embryo proper (Fig. 2.3C and D). At 150 days after pollination, the protoderm has differentiated and the mitotic activity of embryo proper cells has ceased (Fig. 2.3E). During embryo development, the cytoplasm of the embryo proper cells remains dense. Starch granules are the first storage product to appear, and they tend to congregate around the nucleus of the cell. As the seed approaches maturity, the embryo proper is filled with protein and lipid bodies. At seed maturity, an embryo measures about six cells long and five to six cells wide (Fig. 2.3F).

Since *Calanthe* and *Phaius* are relative genus, the pattern of suspensor development of *Calanthe tricarinata* is similar to that of *Phaius tankervilliae* as described by Ye *et al.*[15] In *Calanthe tricarinata*, the suspensor consists of a large, single cell (Fig. 2.3E). During the course of the suspensor development, the suspensor cell begins to enlarge first, then elongates quickly by the process of vacuolation (Fig. 2.3C and D). Finally, the suspensor protrudes beyond the inner integument to make direct and contact with the seed coat cells. At seed maturity, the suspensor eventually degenerates and are compressed (Fig. 2.3F).

The seed coat of this species is derived from the outer integument which has only two cell layers (Fig. 2.3F). The cell wall of the seed coat cells gives a greenish blue color when stained with the TBO stain, indicating the presence of polyphenols and lignin (Fig. 2.4A). The inner

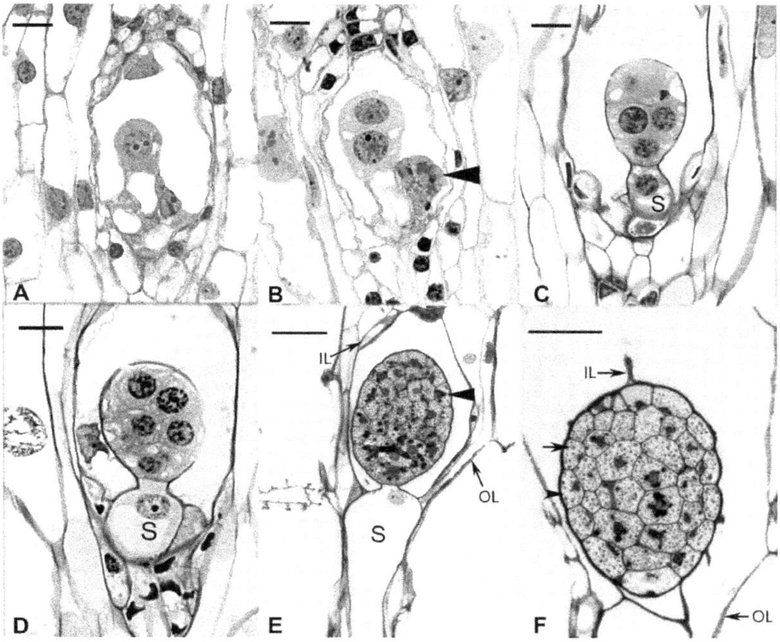

Fig. 2.3. The embryo development of *Calanthe tricarinata*.

(**A**) Light micrograph showing a longitudinal section of a zygote, which is highly polarized with a chalazally located nucleus and a prominent vacuole occupying at the micropylar end. Scale bar = 8 μm.
(**B**) Light micrograph showing a two-celled embryo. The first cell division of the zygote is unequal, resulting in the formation of a smaller terminal cell and a larger basal cell. In this species the endosperm fails to develop. The polar-chalazal complex (*arrowhead*) includes the chalazal nuclei, polar nuclei and one of the male nuclei. Scale bar = 10 μm.
(**C**) Light micrograph showing a six-celled embryo. The cells of the embryo proper have condensed cytoplasm, while the suspensor cell (S) remains highly vacuolated. Scale bar = 8 μm.
(**D**) Light micrograph showing an early globular embryo. The suspensor cell (S) begins to enlarge by vacuolation. Scale bar = 15 μm.
(**E**) Light micrograph showing a globular embryo. At this stage, the mitotic activity of embryo proper cell has ceased. There are starch grains (*arrowhead*) scattered within the embryo proper cells. The cells are also with an abundant deposition of protein and lipid bodies. The suspensor (S) has enlarged and elongated toward the micropylar end of the seed. Both inner (IL) and outer (OL) layer of the seed coat become dehydrated and gradually are compressed into a thin layer. Scale bar = 80 μm.
(**F**) A longitudinal section through a mature seed. As the seed approached maturity, there is a concurrent diminishing of starch granules within the cells. The lipid bodies (*arrowhead*) and protein bodies (*arrow*) of various sizes can be found within the cells of the embryo proper. The suspensor has degenerated at this stage. II, inner layer of the seed coat; OI, outer layer of the seed coat. Scale bar = 60 μm.

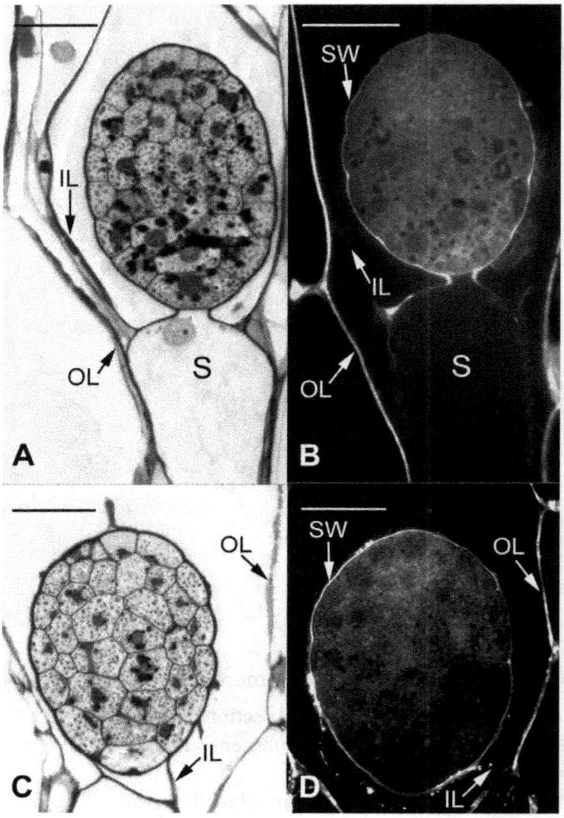

Fig. 2.4. Nile red staining fluorescence micrograph of the embryos of *Calanthe tricarinata*.

(**A**) Light micrograph showing a globular embryo with an enlarged and elongated suspensor (S) toward the micropylar end of the seed. IL, inner layer of the seed coat; OL, outer layer of the seed coat. Scale bar = 70 μm.

(**B**) Nile red staining fluorescence pattern of a developing seed at the stage similar to that in A. The outermost layer of outer layer of the seed coat (OL) and the surface wall (SW) of the embryo proper react positively to the stain, while the inner layer of the seed coat (IL) does not fluoresce brightly. There is a thin layer cuticular substance covering the exposed portion of the suspensor (S). Scale bar = 50 μm.

(**C**) Light micrograph showing a mature seed. The suspensor has degenerated at this stage, and the mature embryo is enveloped by the shriveled seed coat. IL, inner layer of the seed coat; OL, outer layer of the seed coat. Scale bar = 60 μm.

(**D**) Nile red staining fluorescence pattern of a developing seed at the stage similar to that in C. The inner layer of the seed coat (IL) and the outermost layer of the outer layer of the seed coat (OL) reacted positively to the stain. Moreover, the surface wall (SW) of the embryo proper also fluoresced brightly. Scale bar = 50 μm.

integument is composed of two layers of vacuolated parenchyma cells (Fig. 2.3B). The cell wall of the inner integument gives a purple color when stained with the TBO stain, suggesting a lack of a secondary wall. Nile red staining indicates the accumulation of cuticular substances in the outer layer of the seed coat, and the absence of cuticular substance in the inner layer of the seed coat (Fig. 2.4B and D). As the seed matures, the seed coat layers dehydrate and are compressed into thin layers (Fig. 2.4C).

From determination of the endogenous ABA during the course of embryo development, it has been shown that relatively lower levels of ABA (approximate 2 $ng \cdot mg^{-1}$ of fresh weight) are found during early embryo development.[13] Later, the ABA level continues to increase, and is then maintained at a maximal level (approximate 11.6 $ng \cdot mg^{-1}$ of fresh weight) at seed maturity.

2.2.1.3 *Phalaenopsis amabilis var. formosa*

Phalaenopsis amabilis var. *formosa* is an epiphytic, native species of Taiwan. At 60 days after pollination, the zygote takes on an ovoid shape within the small embryo sac (Fig. 2.5A). The nucleus is located toward the chalazal end, and numerous tiny vacuoles are present in the cytoplasm of the zygote. Similar to the majority of orchids, the endosperm fails to develop in this species. Within the endosperm cavity, the endosperm complex eventually degenerates and is absorbed by the developing embryo (Fig. 2.5A and B). The first zygote division is asymmetrical, producing a larger basal cell and a smaller terminal cell (Fig. 2.5B). The basal cell of the two-celled embryo divides further, resulting in the formation of a three-celled embryo (Fig. 2.5C). The upper two cells of the three-celled embryo give rise to the embryo proper, and the larger basal cell becomes the suspensor initial cell. During the early stages of embryo proper formation, oblique cell divisions are commonly observed (Fig. 2.5F). Furthermore, the cells towards the chalazal end divide more frequently than the cells towards the micropylar end, forming two different sizes of cells within an ellipsoidal embryo: smaller cells in the anterior region and larger cells in the posterior end near the suspensor cells (Fig. 2.6B and C). During the early stages of embryo development, no distinct storage products are observed within the cytoplasm of the embryo proper (Fig. 2.6A). After the cells have ceased to divide, starch grains are the first storage product to appear within the embryo proper; the starch grains tend to congregate around the nucleus of the cells (Fig. 2.6B). As the seed approaches maturity, large vacuoles have broken

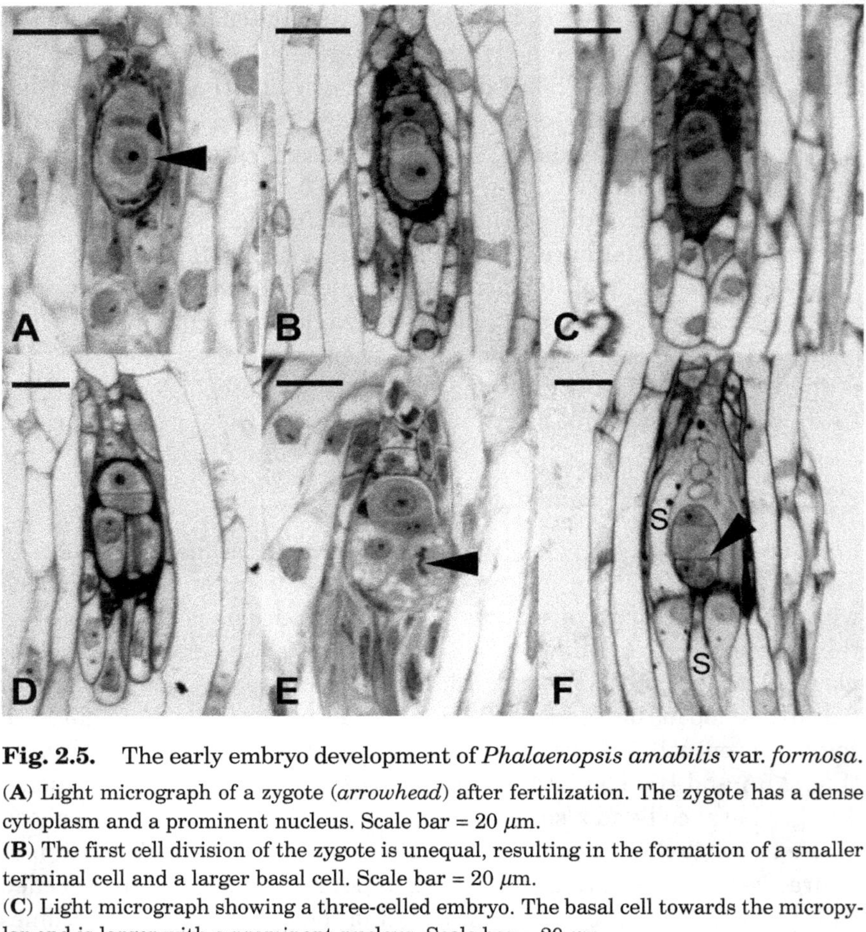

Fig. 2.5. The early embryo development of *Phalaenopsis amabilis* var. *formosa*.

(**A**) Light micrograph of a zygote (*arrowhead*) after fertilization. The zygote has a dense cytoplasm and a prominent nucleus. Scale bar = 20 μm.

(**B**) The first cell division of the zygote is unequal, resulting in the formation of a smaller terminal cell and a larger basal cell. Scale bar = 20 μm.

(**C**) Light micrograph showing a three-celled embryo. The basal cell towards the micropylar end is larger with a prominent nucleus. Scale bar = 20 μm.

(**D**) Light micrograph showing a four-celled embryo. The two cells towards the chalazal end have a dense cytoplasm, whereas the other two cells towards the micropylar end continue to enlarge and elongate. Scale bar = 20 μm.

(**E**) The basal cells toward the micropylar end continue to divide (*arrowhead*), and the daughter cells then differentiate into suspensor cells. In these dividing cells, vacuoles also become more prominent in the cytoplasm. Scale bar = 20 μm.

(**F**) The oblique cell division occurs in the embryo proper (*arrowhead*), which signals the formation of the globular shaped embryo. At the same time, the suspensor cells (S) enlarge and elongate rapidly within the endosperm cavity towards both the micropylar and chalazal ends. Scale bar = 20 μm.

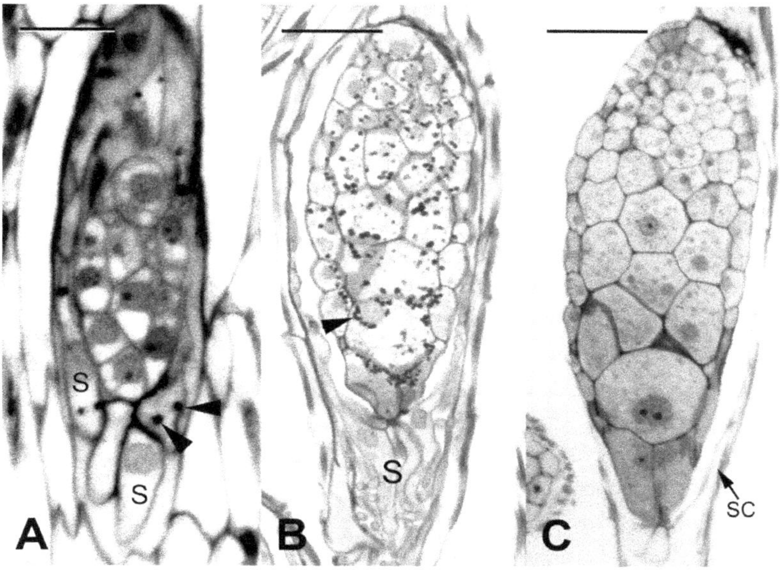

Fig. 2.6. The later stages of embryo development of *Phalaenopsis amabilis* var *formosa*.

(**A**) Light micrograph showing an early globular embryo. The suspensor cells (S) has elongated and surround the embryonic mass. In the Spurr's resin section, it is worthwhile to note that several lipid droplets (*arrowhead*) appeared in the suspensor cells after TBO staining. Scale bar = 25 μm.

(**B**) Light micrograph showing a globular embryo, which consists of two cell types: smaller cells in the anterior region and larger cells in the posterior end near the suspensor. The suspensor cells (S) begin to degenerate at this stage. In the embryo proper cells, starch grains (*arrowhead*) are abundant in the cytoplasm, and tend to congregate around the nucleus. Scale bar = 50 μm.

(**C**) Light micrograph showing a longitudinal section through a mature seed. Several minute protein bodies (*arrowhead*) can be observed within the cells of embryo proper. In the historesin section, lipid is not preserved; the spaces between the protein bodies are occupied by storage lipid bodies. At maturity, the embryo is enclosed by the shriveled seed coat (SC). Scale bar = 50 μm.

down and the starch grains have vanished. At the same time, lipid bodies and protein bodies start to accumulate within the cytoplasm. At seed maturity, protein and lipid bodies are the major storage products within the embryo proper (Fig. 2.6C).

The embryo development of *Phalaenopsis amabilis* var. *formosa* is characterized by a vertical division in the basal cell of a three-celled embryo, resulting in the formation of a ⊥-shaped four-celled embryo (Fig. 2.5D). The two basal cells of the four-celled embryo divide two times and eventually give rise to a plate of eight suspensor cells (Fig. 2.5E). Soon after, the suspensor cells elongate rapidly by vacuolation, and some lipid droplets appear in the cytoplasm of the suspensor cell (Fig. 2.5F). Throughout embryo development, the tubular suspensor cells extend towards the micropylar and the chalazal ends, and surround the embryonic mass (Fig. 2.6A). As the seed approaches maturity, the suspensor cells shrivel, and the remnant of the suspensor could be observed in the mature seed (Fig. 2.6A and B). This pattern of embryo development occurs commonly in the genera of the Vandoid group, such as *Phalaenopsis*, *Rhynchostylis* and *Vanda*.[16] The oblique division of the proembryo is characteristic of the *Cymbidium* type, suggesting that it is a programmed process that results in the formation of the suspensor structure[17]; in *Phalaenopsis*, the formation of a ⊥-shaped four-celled embryo may be the trait of the Vandoid group that has the eight-celled filamentous suspensor.

The mature seed coat is derived from the outer integument, which is two-three cells thick. The cells of the inner integument gradually collapse at the proembryo stage (Fig. 2.5F). The thickened radial walls of the outermost layer of the seed coat stain greenish blue with the TBO stain, indicating the presence of secondary walls (Fig. 2.7A). Autofluorescence of the seed coat indicates that lignin is present. Staining using Nile red indicates that cuticular materials are also present in the secondary walls (Fig. 2.7B). At maturity, the cells of seed coat become dehydrated and are compressed into a thin layer (Fig. 2.6C).

2.3 Integuments and the Seed Coat

At the time of seed dispersal, the mature embryo of orchids is protected by a thin seed coat. The seed coat is derived from the maternal tissues — the integuments of the ovules. In *Cymbidium sinense*,[17] the seed coat is derived from both the inner and outer integuments. The inner integument of some species degenerates in the early stages of embryo development, and the seed coat is derived from the outer integument, such as *Calypso bulbosa*[18] and *Phalaenopsis*.[19] Generally, the inner layers

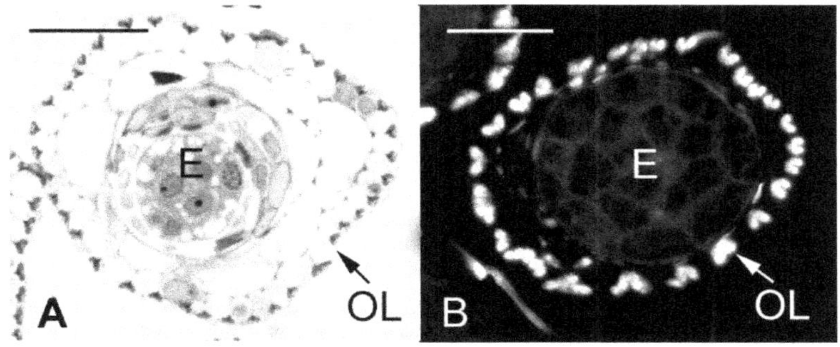

Fig. 2.7. Nile red staining fluorescence micrograph of the developing embryo of *Phalaenopsis amabilis* var. *formosa*.

(**A**) Light micrograph showing a cross section of a globular embryo of *Phalaenopsis amabilis* var. *formosa*. Secondary wall was present in the radial walls of the outer layer of the seed coat. E, embryo proper; OL, outer layer of the seed coat. Scale bar = 50 μm.
(**B**) Nile red staining fluorescence micrograph of a developing seed at the stage similar to that in A. The radial walls of outer layer of the seed coat (OL) and the surface wall of the embryo proper react positively to the stain. E, embryo proper; OL, outer layer of the seed coat. Scale bar = 50 μm.

of the seed coat is composed of two layers of vacuolated parenchyma cells (Figs. 2.2A and 2.3B). In *Calypso bulbosa*[18] and *Epidendrum ibaguense* (Fig. 2.8), the inner layers of the seed coat that originated from the inner integument gives a strong positive protein staining during the ovule and early embryo developmental stages, suggesting that these cell layers may play a role as the nutrient supplier during these stages of development. In the outer layers of the seed coat, the cells are usually vacuolated. In many species, such as *Cypripedium formosanum*, starch grains can be found in the cells of the outer layers of the seed coat at the early stage of embryo development (Fig. 2.2A), and the starch grains disappear as the seed approaches maturity (Fig. 2.2F). As the seed matured, the seed coat cells dehydrate and are compressed into a thin layer, which envelopes the embryo.

Although the seed coat of an orchid seed looks simple, the developmental pattern and the addition of secondary walls are diverse among the species. During seed development, cuticular materials, phenolic substances, and lignin may accumulate differently in the seeds coat. In *Calanthe tricarinata* (Fig. 2.4), *Cymbidium sinense*[17] and *Cypripedium formosanum*,[12] cuticular material as well as phenolic compounds are

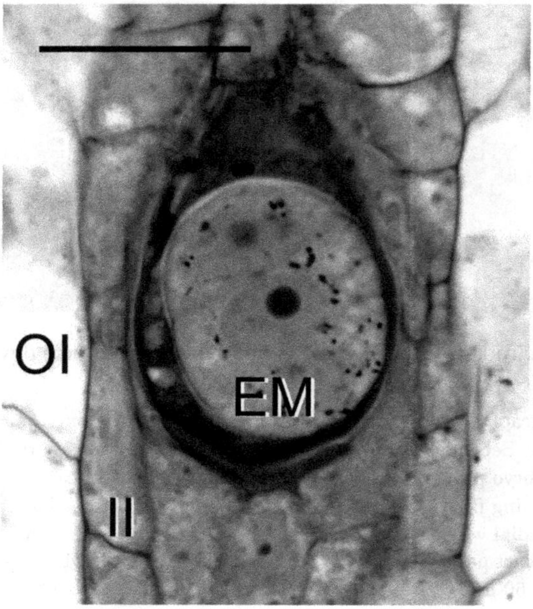

Fig. 2.8. Light micrograph showing a two-celled embryo of *Epidendrum ibaguense*. The inner integument cells gives a strong positive protein staining, Amido black 10B. Scale bar = 20 μm.

present in the outermost layer of the seed coat, and persists through seed maturation. In *Paphiopedilum delenatii*, the cuticular material is present over the innermost walls of the inner layer of the seed coat that encloses the endosperm cavity at the early globular stage, while the outer layers lack the cuticular material.[20] From previous reports, mature embryos of some terrestrial species are covered by a unique structure which is known as the "inner seed coat" or the "carapace."[21,22] The embryo with thick and complete "carapace" is hard to germinate, such as *Cephalanthera*,[11,22] *Cypripedium*[21] and *Epipactis*.[22] Based on a series histological observation of developing seeds, the carapace is derived from the inner integument that envelops the embryo tightly as the seed matures.[12] In *Phalaenopsis amabilis* var. *formosa*, an easy-to-germinate epiphytic species, the presence of lignin and cuticular material are discontinuous, accumulating mainly in the radial walls of the outermost layer of the seed coat. This feature may allow water and nutrients access to the embryo for germination (Fig. 2.7A and B). In *Cypripedium formosanum*, Nile red staining indicates the accumulation of cuticular substance of the carapace.[12] Ultrastructural observation also confirms the presence of an electron dense layer in the surface wall of the carapace (unpublished data). Seed germination of the terrestrial species is usually difficult. Treating mature seeds by calcium hypochlorite, sodium hypochlorite, or ultrasound may scarify and break the firm covering of the embryo, and thus improve seed germination.[13,23–25]

2.4 A General Discussion on Features of Orchid Embryo Development

2.4.1 *The lack of a clearly histodifferentiation pattern*

One of the most striking features of embryo development is the lack of a clear histodifferentiation pattern (Figs. 2.2F and 2.3F). As illustrated in the above case histories, the embryo proper lacks defined tissues such as the apical meristems, cotyledon and primary tissue system. Although a protoderm is present, its organization is not as well defined as in other flowering plants such *Brassica napus*.[17] In an easier and faster germinating species, *Phalaenopsis amabilis* var. *formosa*, a gradient of cell size is observed, with smaller cells occupying the future shoot pole (Fig. 2.6C). This difference in cell size may represent different developmental potential and may be a reflection of biological differentiation of cells within the embryo proper. Is the absence of a distinct histodifferentiation pattern due to the absent of the nutritive tissue, the endosperm? In flowering plants, the endosperm is an integral part of seed and embryo development. Abnormalities in endosperm development often result in embryo abortion.[26] The failure of endosperm development may cause abnormal differentiation and starvation of the developing embryos.[27,28] The endosperm is the food storage site for the developing embryo in most flowering plants, and the lack of an endosperm in orchids may have prevented further histodifferentiation of the globular embryo. Moreover, the absence of an endosperm results in an "empty" cavity. Since the seed coat of orchids is thin and without the nutritive endosperm tissue serving as buffer, the developing embryo may be subject to water stress. In *Cypripedium formosanum*[14] and *Calanthe tricarinata*,[13] the speedy increase of the ABA content is found as the seeds approach maturity. The rapid increase in the ABA content may be an indication of the unusually "dry" embryonic environment. It is well known that a majority of the developmental process cannot tolerate the lowering of water content and water stress. Mitotic divisions and DNA synthesis are both sensitive to water stress and increasing ABA levels.[29,30] In maize, water deficit decreases the rate of endosperm cell division and inhibits DNA synthesis.[31] If this is indeed the case, the mitotic activity becomes arrested early during the orchid embryo development. Hence, further histodifferentiation simply cannot occur.

2.4.2 *Presence of a cuticle over the embryo proper*

All exposed plant surfaces have a cuticle. Rodkiewicz *et al.*[32,33] clearly establish that the embryo proper in flowering plants has a distinct cuticle covering its surface, while a cuticle is absent from the suspensor. In orchid embryos, as shown in Section 2.2.2, a clearly defined cuticle is present over the entire embryo proper surface (Fig. 2.4B and D). In *Cymbidium sinense*,[17] *Paphiopedilum delenatii*,[20] and *Phalaenopsis amabilis* var. *formosa*,[19] a cuticle can be detected as early as the globular stage of development. The formation of a cuticle at such as an early stage supports the notion that the embryonic environment may be "dry." In plants, a cuticle functions to prevent water loss from the surface of the plant body. The cuticle covering the orchid embryo proper most likely serves to protect it from premature desiccation. The negative side of having a cuticle is that it may retard or impede nutrient absorption directly from the surface of the embryo proper.

2.4.3 *The absence of a cuticle in the suspensor cell wall*

Histochemical studies using Nile red staining reveal that cuticular substances are absent from the suspensor cell wall.[17,20] This observation clearly indicates structural and functional differences between the two parts of the embryo. In orchids as well as in a majority of flowering plants, the suspensor is usually embedded within the integumentary tissues and is not "exposed" (Figs. 2.3E and 2.6B). This may be the primary reason why a cuticle is absent from the suspensor. The consequence of not having a cuticle is important in facilitating nutrient uptake from the maternal tissues as the embryo is attached to the seed coat via the suspensor.

The suspensor is a short-lived embryonic organ. Because of its precocious development during early embryogeny, together with its structural and physiological characteristics, the suspensor is believed to be essential to early embryo development.[34] One of the important functions of the suspensor is to serve as a channel of nutrient uptake for the developing embryo as it connects the embryo proper to the maternal tissues.[34] In the study of plant cell biology, the staining characteristics of a cell wall indirectly indicate the structure and function of the cell of interest. When suspensors are stained with toluidine blue O, the cell walls give a purple color (Fig. 2.4A). The walls stain negatively towards lipid stains such as Nile red and negatively toward the phloroglucinol-HCl stain. These stains clearly indicate the cell wall is "primary" in nature and the absence of a cuticle and lignin in the wall (Fig. 2.4B). Since the primary cell wall allows for apoplastic movement of water and dissolved solutes, the suspensor cell wall can form an apoplastic continuum with cells of the maternal tissues, i.e. the seed coat. Albeit not necessary an efficient translocation process, the apoplastic continuum ensures nutrient flow directly from the maternal tissue to the embryo proper via the suspensor.

2.4.4 *A suspensor having a "transfer cell morphology"*

It is well established that orchid suspensors have a varied morphology (Table 2.1); their nutritive function is usually inferred from their pattern of growth. At the light microscope level, orchid suspensor cells are usually highly vacuolated with no obvious structural specialization detected. The highly vacuolated cells together with the absence of structural specializations such as wall ingrowths suggest that the orchid suspensor may not have specialized functions similar to that reported in other flowering plants, such as the *Phaseolus* species.[35,36] Ultrastructural studies concerning orchid suspensors are rare. As indicated above, in our recent ultrastructural studies, we noted that the *Paphiopedilum delenatii* suspensor indeed takes on a "transfer cell" morphology, indicating that the suspensor can play an important role in embryo development.[20]

The transfer cells function as the short-distance transporter for solutes across cells through wall ingrowths. In developing *Vicia faba* seeds, genes involved in plasma membrane sucrose transport (H^+/sucrose symporter gene, sucrose binding protein gene, and H+-ATPase gene) are found to express in the wall ingrowths.[37] Furthermore, H^+/amino acid co-transporter genes are found to express predominantly in the transfer cells of pea seeds.[38] These observations provide strong evidence that transfer cells play an important role in the uptake of the solutes. Our recent ultrastructural study clearly indicates that the suspensor of *Paphiopedilum delenatii* takes on a "transfer cell" morphology[20]; (Fig. 2.9A and B). Although there is no experimental data indicating that the orchid suspensor cells have similar biochemical functions as indicated above, having a transfer cell morphology strongly suggests that similar biochemical properties and functions are possible. In *Paphiopedilum delenatii*, wall ingrowths begin to appear when the embryo reaches the six-cell stage.[20] The wall ingrowths are strategically located on the side walls abutting the maternal seed coat. The increase in the surface area of the suspensor cell supports the idea of enhancing nutrient uptake at this location. The early formation of wall ingrowths suggests that the ingrowths are important and have unique functions to play during the early stages of orchid embryo development. Additional ultrastructural

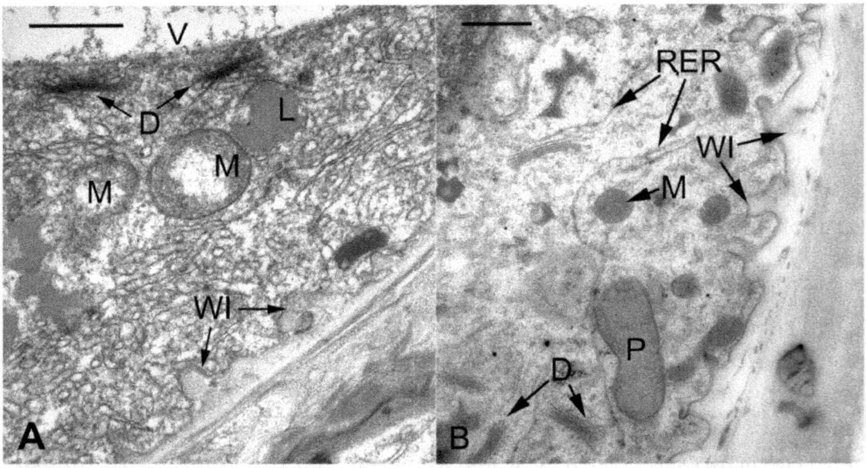

Fig. 2.9. Electron micrographs of the suspensor development of *Paphiopedilum delenatii* embryo.

(**A**) Electron micrograph of the micropylar end of the suspensor at the proembryo stage, showing the formation of the wall ingrowth (WI) along the basal wall. The cytoplasm is filled with the tubular SER system. D = dictyosomes; L = lipid body; M = mitochondrion; V = vacuole. Scale bar = 1 μm.

(**B**) Electron micrograph showing the micropylar end of the fully developed suspensor at the globular stage. At this stage, the wall ingrowths (WI) are well developed; the rough ER (RER) begins to appear, which is in the form of long strands. The plastids (P), dictyosomes (D) and mitochondria (M) are plentiful in the cytoplasm. Scale bar = 1 μm.

studies are needed to determine whether structural specializations are present in other orchid suspensor systems.

2.4.5 *The accumulation of high levels of ABA in mature seeds of some terrestrial species*

ABA is essential in regulating seed embryonic maturation, dormancy and germination.[39] In orchids, there are only a few reports on the endogenous ABA concentrations in developing seeds. In *Epipactis helleborine*, it has been shown that the endogenous ABA level in mature seeds is 14-fold greater than that in immature seeds.[40] In recent studies on *Calanthe tricarinata*[13] and *Cypripedium formosanum*,[14] as the seeds approach maturity, there is a rapid increase in the ABA concentration (Fig. 2.4). In cereals, the ABA concentration is low in early developing seeds, reaches the highest concentration during mid-development, and then declines as the seed matures.[41–43] In the developing seeds of *Calanthe tricarinata* and *Cypripedium formosanum*, the ABA concentration does not decrease, but continues to increase as the seeds approach maturity. Furthermore, *in vitro* germination decreases sharply as the seeds approach maturity; and this coincides with increasing ABA levels, and decreasing water content. These results suggest that ABA may regulate the processes of storage protein synthesis, acquisition of desiccation tolerance, and prevention of germination in orchid seeds.

2.5 Conclusion and Perspectives

In Taiwan, a number of orchids, such as *Cymbidium*, *Paphiopedilum*, *Phalaenopsis* and *Oncidium* have become important floricultural crops. In addition, there are numerous native species with ornamental value. Asymbiotic germination is a useful and popular technique in obtaining seedlings for nursery culture, and to conserve our natural resources. Although most orchids can germinate successfully on a defined medium, asymbiotic seed germination is often intricate for some terrestrial orchids. A detailed knowledge of embryo development and its structure will facilitate our success in asymbiotic seed germination by selecting the proper age of developing seeds for *in vitro* germination. Our knowledge concerning the timing of the cessation of mitotic activity, appearance of the cuticle in the embryo, and presence or absence of specialization in cells of the embryo and surrounding maternal tissues will enable us to determine the most opportune time for the asymbiotic germination of immature seeds. The structure of the seed coat, thickness of the cuticle surrounding the embryo proper, cell number and presence or absence of a distinct cell size gradient also allow us to estimate the ease of symbiotic and asymbiotic germination of mature seeds.

References

1. Arditti J. (1992) *Fundamentals of Orchid Biology*. John Wiley and Sons, Inc, New York.
2. Ye, XL, Yeung EC, Zee SY. (2002) Sperm movement during double fertilization of a flowering plant, *Phaius tankervilliae*. *Planta* **215:**60–66.
3. Molvray M, Chase MW. (1999). Seed morphology. In: Pridgeon AM, Cribb PJ, Chase MW, Rasmussen FN (eds.), *Genera Orchidacearum*, Vol. 1. Oxford University Press, Oxford, UK, pp. 59–66.
4. Yam TW, Yeung EC, Ye XL, *et al.* (2002) Orchid embryos. In: Kull T, Arditti J (eds.), *Orchid Biology: Reviews and Perspectives*, 8th edn. Kluwer, Dordrecht, pp. 287–385.
5. Yam TW, Nair H, Hew CS, Arditti J. (2002) Orchid seeds and their germination: an historical account. In: Kull T, Arditti J (eds.). *Orchid Biology: Reviews and Perspectives*, 8th edn. Kluwer, Dordrecht, pp. 387–504.
6. Knudson L. (1922) Nonsymbiotic germination of orchid seeds. *Bot Gaz* **73:**1–25.
7. Clements MA. (1999) Embryology. In: Pridgeon AM, Cribb PJ, Chase MW, Rasmussen FN (eds.), *Genera Orchidacearum*, Vol. 1. Oxford University Press, Oxford, UK, pp. 38–66.
8. Batygina TB, Bragina EA, Vasilyeva VE. (2003) The reproductive system and germination in orchids. *Acta Biol Cracov Series Botanica* **45:**21–34.
9. Andronova EV. (2006) Embryogenesis in *Orchidaceae*. In: Batygina TB (ed.), *Embryology of Flowering Plants, Vol. 2 Seed.* Science Publishers, New Hampshire, pp. 355–363.
10. Rasmussen HN. (1995) *Terrestrial Orchids from Seed to Mycotrophic Plants*. Cambridge University Press, Cambridge.
11. Yamazaki J, Miyoshi K. (2006) *In vitro* asymbiotic germination of immature seed and formation of protocorm by *Cephalanthera falcata* (Orchidaceae). *Ann Bot* **98:**1197–1206.

12. Lee YI, Lee N, Yeung EC, Chung MC. (2005) Embryo development of *Cypripedium formosanum* in relation to seed germination *in vitro*. *J Amer Soc Hort Sci* **130:**747–753.
13. Lee YI, Lu CF, Chung MC, *et al.* (2007) Developmental changes in endogenous abscisic acid concentrations and asymbiotic seed germination of a terrestrial orchid, *Calanthe tricarinata* Lindl. *J Amer Soc Hort Sci* **132:**246–252.
14. Lee YI. (2003) Growth periodicity, changes of endogenous abscisic acid during embryogenesis, and *in vitro* propagation of *Cypripedium formosanum* Hay, Ph.D dissertation. National Taiwan University, Taipei, Taiwan, ROC.
15. Ye XL, Zee SY, Yeung EC. (1997) Suspensor development in the nun orchid, *Phaius tankervilliae*. *Intern J Plant Sci* **158:**704–712.
16. Swamy BGL. (1949) Embryological studies in the *Orchidaceae*. II. Embryology. *Am Midland Naturalist* **41:**202–232.
17. Yeung EC, Zee SY, Ye XL. (1996) Embryology of *Cymbidium sinense*: Embryo development. *Ann Bot* **78:**105–110.
18. Yeung EC, Law SK. (1992) Embryology of *Calypso bulbosa*. II. Embryo development. *Can J Bot* **70:**461–468.
19. Lee YI, Yeung EC, Lee N, Chung MC. (2007) Embryology of *Phalaenopsis amabilis* var. *formosa*: Embryo development. *Botanical Study* (submitted).
20. Lee YI, Yeung EC, Lee N, Chung MC. (2006) Embryo development in the lady's slipper orchid, *Paphiopedilum delenatii* with emphases on the ultrastructure of the suspensor. *Ann Bot* **98:**1311–1319.
21. Carlson MC. (1940) Formation of the seed of *Cypripedium parviflorum*. *Bot Gaz* **102:**295–301.
22. Veyret Y. (1969) La structure des semences des orchidaceae et leur aptitude a la germination *in vitro* en cultures pures. Musee d'Histoire Naturelle de Paris, *Travaux du Laboratoire La Jaysinia* **3:**89–98.
23. Van Waes JM, Debergh PC. (1986) Adaption of the tetrazolium method for testing the seed viability and scanning electron microscopy of some Western European orchids. *Physiol Plant* **66:**435–442.
24. Van Waes JM, Debergh PC. (1986) *In vitro* germination of some Western European orchids. *Physiol Plant* **67:**253–261.
25. Miyoshi K, Mii M. (1988) Ultrasonic treatment for enhancing seed germination of terrestrial orchid, *Calanthe discolor*, in asymbiotic culture. *Scientia Hort* **35:**127–130.
26. White DWR, Williams E. (1976) Early seed development after crossing of *Trifolium semipilosum* and *T. repens*. *NZ J Bot* **74:**161–168.
27. Lester RN, Kang JH. (1988) Embryo and endosperm function and failure in *Solanum* species and hybrids. *Ann Bot* **82:**445–453.

28. Niroula RK, Bimb HP, Sah BP. (2006) Interspecific hybrids of buckwheat (*Fagopyrum* spp.) regenerated through embryo rescue. *Scientific World* **4:**74–77.
29. Levi M, Brusa P, Chiatante D, Sparvoli E. (1993) Cell cycle reactivation in cultured pea embryo axes: Effect of abscisic acid. *In Vitro Cell Dev Biol Plant* **29P:**47–50.
30. Liu Y, Bergervoet JHW, Ric De Vos CH, *et al.* (1994) Nuclear replication activities during imbibition of abscisic acid- and gibberellin-deficient tomato (*Lycopersicon esculentum* Mill.) seeds. *Planta* **194:**368–373.
31. Setter TL, Flannigan BA. (2001) Water deficit inhibits cell division and expression of transcripts involved in cell proliferation and endoreduplication in maize endosperm. *J Expt Bot* **52:**1401–1408.
32. Rodkiewicz B, Fyk B, Szczuka E. (1994) Chlorophyll and cutin in early embryogenesis in *Capsella, Arabidopsis*, and *Stellaria* investigated by fluorescence microscopy. *Sex Plant Reprod* **7:**287–289.
33. Rodkiewicz B, Szczuka E. (2006) Cuticle of developing embryo. In: Batygina TB (ed.), *Embryology of Flowering Plants, Vol. 2, Seed.* Science Publishers, New Hampshire, pp. 373–374.
34. Yeung EC, Meink DW. (1993) Embryogenesis in angiosperms: Development of the suspensor. *Plant Cell* **5:**1371–1381.
35. Yeung EC, Clutter ME. (1978) Embryogeny of *Phaseolus coccineus*: Growth and microanatomy. *Protoplasma* **94:**19–40.
36. Yeung EC, Clutter ME. (1979) Embryology of *Phaseolus coccineus*: The ultrastructure and development of the suspensor. *Can J Bot* **57:**120–136.
37. Harrington GN, Franceschi VR, Offler CE, *et al.* (1997) Cell specific expression of three genes involved in plasma membrane sucrose transport in developing *Vicia faba* seed. *Protoplasma* **197:**160–173.
38. Tegeder M, Offler CE, Frommer WB, Patrick JW. (2000) Amino acid transporters are localized to transfer cells of developing pea seeds. *Plant Physiol* **122:**319–325.
39. Bewley JD, Black M. (1985) *Seeds: Physiology of the Development and Germination*. Plenum Press, New York, USA.
40. Van der Kinderen G. (1987) Abscisic acid in terrestrial orchid seeds: A possible impact on their germination. *Lindleyana* **2:**84–87.
41. Goldbach H, Michael G. (1976) Abscisic acid content of barley grains during ripening as affected by temperature and variety. *Crop Sci* **16:**797–799.
42. Kawakami N, Miyake Y, Noda K. (1997) ABA insensitivity and low ABA levels during seed development of non-dormant wheat mutants. *J Expt Bot* **48:**1415–1421.
43. King RW. (1976) Abscisic acid in developing wheat grains and its relationship to grain growth and maturation. *Planta* **132:**43–51.

Chapter 3

In vitro Morphogenesis and Micro-Propagation of Orchids

*Wei-Chin Chang**

Plant biotechnology, especially *in vitro* regeneration and flowering, cell biology, DNA manipulation, and biochemical engineering, is reshaping orchid research in four major areas: 1) benefiting micro-propagation and transgenic research with findings on the totipotency and regeneration ability through shoot-bud formation and somatic embryogenesis from callus, direct somatic embryogenesis from explants and thin-section cultures of leaves and roots, and even shoot-bud formation of suspension cells of several major commercial orchids; 2) active research into the dissection of genes responsible for controlling growth, meristem functioning, and flowering of orchids; 3) successful application of molecular genetics and plant transformation under laboratory conditions for protecting commercial orchids against biotic stress; 4) production of specialty biochemicals and pharmaceuticals. These are all good starts, but more devotion and support are needed for further research into both basic and practical aspects. This chapter will include findings on: 1) direct somatic embryogenesis of leaf explants of *Oncidium*, *Phalaenopsis*, and *Dendrobium*; 2) direct shoot-bud formation from leaf explants and regeneration from suspension cells of *Paphiopedilum*; 3) *in vitro* flowering of callus-derived somatic embryos and plantlets of *Cymbidium*, *Dendrobium*, and *Phalaenopsis*; and 4) disease-resistant *Dendrobium* and *Oncidium* through genetic transformation. For the challenges ahead in orchid biotechnology and its practical application, I strongly encourage integrating modern

Institute of Plant and Microbial Biology, Academia Sinica, Taipei, Taiwan.

technologies with classical breeding for the future success of the commercial orchid industry.

3.1 Introduction

This chapter is meant to be a personal view on the current advances in the micro-propagation of and transgenic approach to commercial orchids, not a review of all the current and past literature on the subject. I knew little about orchids, especially about their *in vitro* culture, 10 years ago. However, during the last 10 years, orchids have consumed my interest and my desire to learn more about exotic plants.

When I finished my training at the University of California, Riverside, in 1972 and returned to Taipei to build a laboratory for working on *in vitro* morphogenesis at Academia Sinica, orchids were not the plants we chose as our early experimental material. Many colleagues in the academic community and the industry had worked on orchids for years and produced many publications on orchid tissue culture. Working on a variety of plants, we eventually focused on ginseng (*Panax ginseng* C. A Meyer), a popular herb among the Chinese community. We learned a lot about somatic embryogenesis and plant regeneration on ginseng callus culture[1] and found a way to reduce its juvenility by inducing the callus-derived embryo, which flowered in a few months under a definite condition.[2] Those experiences helped a lot in our later attempts at somatic embryogenesis and *in vitro* flowering in cultures of bamboo[3,4] and orchids,[5,6] the latter of which I will focus on in this chapter.

3.2 *Cymbidium*

Cymbidium was the first orchid genus we learned to work on *in vitro* in the mid-1990s. Chen Chang, a Ph.D. student in my lab, tried to induce callus from various tissue explants of *Cymbidium ensifolium* var. misericors. The explants included roots, leaves, and pseudo-bulbs of 7- to 8-cm seedlings derived from seed-derived rhizomes and the rhizomes themselves. Chang eventually found that a medium of 1/2

Murashige and Skoog (MS) plus (2,4-dichlorophoxyacetic acid (2,4-D) (10 mg/l) and thioridazine TDZ (0.1 mg/l) was effective for callusing from explants of pseudo-bulbs and rhizomes. However, callusing at that time was a slow process: it took 12- to 18-months to obtain reliable callus for subculture. Those calli, sub-cultured on a 1/2 MS plus a smaller amount of 2,4-D (3.3 mg/l) and the same amount of TDZ (0.3 mg/l) demonstrated their totipotency through three different morphogenetic routes, including granular embryoid-like structures. Plants were regenerated from these calli through two morphogenetic routes: 1) rhizomes produced on basal medium plus 5 mg/l benzyaminopurine (BA), from which plantlets proliferated on basal medium plus 2 mg/l BA; and 2) embryoids formed on basal medium plus 5 mg/l BA or 1 mg/l TDZ, from which rhizomes developed on 1/10 basal medium plus coconut milk (150 ml/l), with lateral buds proliferating and eventually developing into plantlets on basal medium plus 2 mg/l BA (Fig. 3.1).[5]

We learned a lot about *in vitro* orchid culture from the early work on *Cymbidium*, especially the important role of TDZ in callusing and embryogenesis. Were those protocols useful for propagation of *Cymbidium* on a large scale? We demonstrated that one piece of rhizome produced seven shoot buds in 45 days cultured in agitated 1/2 MS liquid medium plus appropriate amounts of 2iP, TDZ, BA and naphthaleneacetic acid (NAA).[7] Plantlets were then transferred onto with the same Gelrite-gelled basal medium supplemented with banana pulp for five months before being put into pots in the greenhouse. The scale was satisfactory for large-scale propagation, but still time consuming.

Similar work on callusing from protocorm-like bodies (PLBs) of *Cymbidium* Twilight Moon "Day Light" was reported by Huan *et al.*[8] The authors obtained embryogenic calli from longitudinally bisected segments of PLBs on modified Vacin and Went medium plus NAA or 2,4-D alone or in combination with TDZ within one month. The medium containing the combination of 0.1 mg/l NAA and 0.01 mg/l TDZ was optimal for callus formation and also good for callus subculture. Embryos/PLBs formed when callus was transferred to the basal medium devoid of hormonal additives. These callus-derived PLBs converted into normal plants with well-developed shoots and roots on the basal medium without hormones after about four months. Proper acclimatization in a green house gave with 100% survival rate and

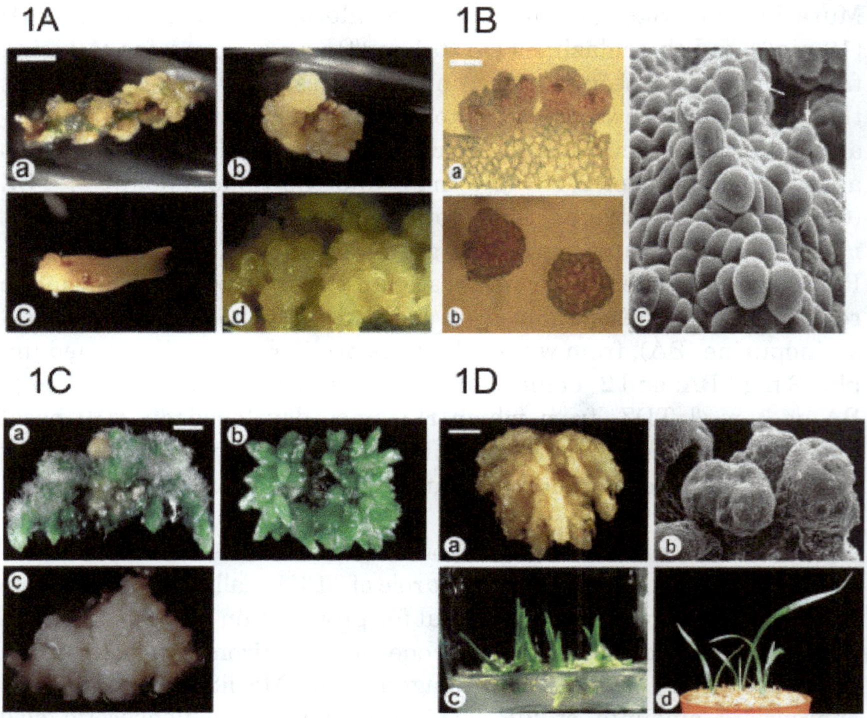

Fig. 3.1. *In vitro* morphogenesis of *Cymbidium ensifolium* var. misericors.

3.1A Callus from lateral buds of rhizome explant (a); callus from a pseudobulb explant (b); callus form a root explant (c), callus from rhizome, after 3 subcultures (d).

3.1B Morphogenesis and structure of granular embryoids: longitudinal section of cluster of white granular embryoids (a); cross-section of white granular embryoids (b); secondary granular embryoids (arrows) proliferating from the surface of a granule (c).

3.1C Morphogenesis of subcultured rhizome calli: rhizomes (a); shoot buds (b), white granular embryoids (c).

3.1D Plantlet regeneration: granular embryoids developing into rhizomes (a); early germination stage of granular embryoids (b); rhizome-derived shoot buds (c); potted plantlets derived from granular embryoids (d).

(*From* Chang and Chang (1998). *Plant Cell Rep* **17**:251–255.)

demonstrated the efficiency of the method for micro-propagation of *Cymbidium*.

We even found a repeatable way to encourage *in vitro* flowering of the sub-cultured callus-derived embryos and plantlets. More intensive studies on *in vitro* flowering led to a report on the role of cytokinins in flowering (Fig. 3.2).[6] Among the eight cytokinins tested, only TDZ, 2iP, and BA induced *de novo* flowering of rhizomes. Callus-derived rhizomes produced flowers precociously within 100 days on 1/2 MS containing TDZ (33.3–10 μM) or 2iP (10–33 μM) combined with 1.5 μM NAA. The flowers we made bloomed for two weeks *in vitro* and were physically normal, with high pollen viability (Fig. 3.2). This was a promising result on the attempt to donate off-season pollen for other parents in terrestrial *Cymbidium* breeding programs.

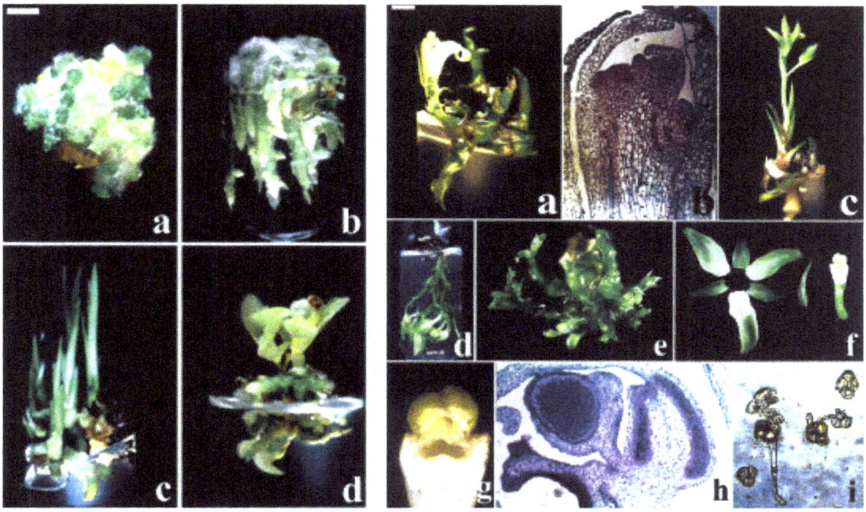

2A 2B

Fig. 3.2. *In vitro* flowering of *Cymbidium ensifolium* var. misericors.

3.2A Morphogenesis of callus-derived rhizomes *in vitro*: rhizome explant for *in vitro* flowering study (a); proliferated rhizomes on basal medium after 100 days of culture (b); shoot formed on basal medium supplemented with TDZ and NAA for 100 days (c); inflorescences formed on basal medium supplemented with BA and NAA after 100 days (d).

3.2B *In vitro* floral morphology of *Cymbidium ensifolium* var. misericors: undersized inflorescences derived from the apex and lateral buds of the rhizome (a); apex of young inflorescence that contains several floret primordia (b); erect raceme inflorescences (c); gravitropic panicle inflorescences (d), inflorescence clumps (e); normal flower with three sepals, 3 petals, and column (f); column with anther and stigma (g); longitudinal section of a column showing anther stigma and pollen (h); germinated pollen (I).

(*From* Chang and Chang (2003). *Plant Growth Reg* **39**:217–221.)

We admired a fine piece of work on *in vitro* flowering of *Cymbidium* published in 1999 by Kstenyuk *et al.*[9] The authors used a combined treatment of cytokinins (BA), restricted nitrogen supply with phosphorus enrichment, and root excision to induce the shoots of 90-day-old rhizome-derived plantlets to flower. A few plants even produce fruit when self-pollinated.

Chang *et al.*[10] also checked the early morpghogenesis of seed germination of *Cymbidium dayanum* reichb.

3.3 *Oncidium*

Oncidium were usually clonally propagated from the dormant buds on flower stalks. The method is quite effective and may not need other protocols to replace it on an industrial scale. Since 1999, we have released a series of reports on direct somatic embryogenesis for leaves and calli derived from a variety of vegetative and reproductive tissues. Segments taken from young leaves of *Oncidium* "Grower Ramsey" produced clusters of somatic embryos directly from epidermal and mesophyll cells of leaf tips and wound surfaces without any intervening callus within one month on culture with Gelrite-gelled 1/2 MS supplemented with low dosage (0.3–1 mg/l) TDZ (Fig. 3.3).[11] Sub-cultures of these embryo clusters produced more embryos and subsequent plantlet formation on the same medium. Similar data were found with *Oncidium* "Sweet Sugar,"[12] but requirements for cytokinin and medium composition for embryo formation were not exactly the same as in "Gower Ramsey." The effects of different cytokinins and auxins and their combinations have been tested.[13] Cytokinins played a promotive role on direct embryo formation, while auxins in general retarded the process. Embryo formation was significantly affected by explant position.[14] Leaf tip segments had a significantly higher embryogenic response than other segments of leaves. Adaxial-side-up orientation significantly promoted embryogenesis as compared to abaxial-side-up orientation. The best response on direct embryo formation was obtained on modified 1/2 MS medium supplemented with 10–20 g/l sucrose, 170 mg/l NaH_2PO_4, and 0.5 g/l peptone. We also found that three ethylene synthesis inhibitors, 1-aminocyclopropane-1-carbocylic acid (ACC), silver nitrate, and cobalt chloride enhanced embryo

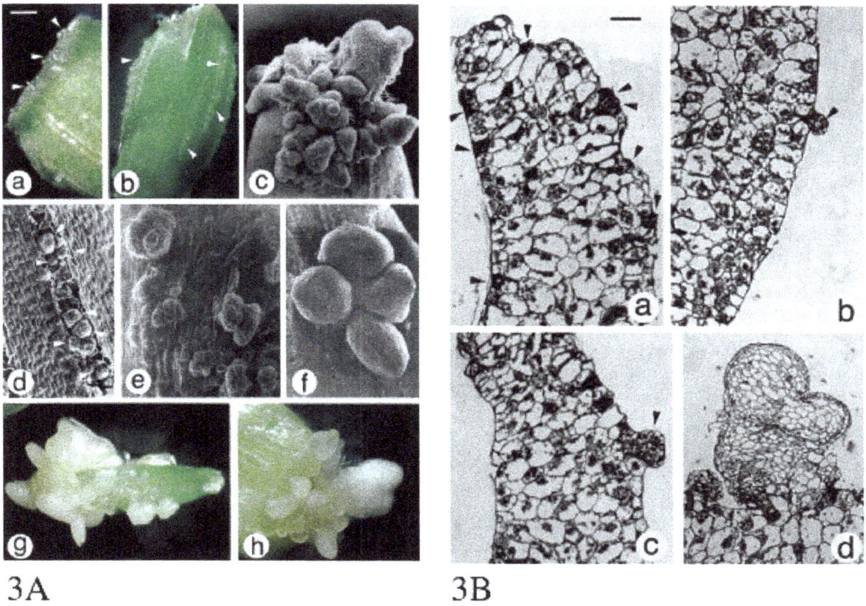

Fig. 3.3. Direct somatic embryogenesis of *Oncidium*.

3.3A *In vitro* morphogenesis of leaf explants on 1/2 MS basal medium plus TDZ: embyogenic nodular mass protruding from the wound surfaces (a); embryogenic nodular mass on epidermal cell layers of leaf explant (b); cluster of embryos on tip of leaf explant (c); small globular embryo formed form the mesophyll cells under the leaf surface (d); small PLBs eventually formed from embryos (e); cluster of 4 young PLBs from embryos (f); PLBs developed on the surface of a piece of leaf explant (g); tip of a leaf explant bearing a cluster of embryos.

3.3B Morphogenesis of embryos and PLBs: cross section of a leaf explant showing embryonic cell of epidermal layer with densely stained cytoplasm and nuclei (a); a young embryo with approximately 10 cells protruding from the epidermal layer of a leaf explant (b); a developing embryo with more embryonic cells (c); well-developed embryo or young PLB protruding into the adjacent surface (d).

(*From* Chen *et al*. (1999) *Plant Cell Rep* **19**:143–149.)

formation.[15] Gibberellic acid (GA_3), cycocel, and paclobutrazol, in general, retarded embryo formation, while ancymidol in low concentration had a promotive effect.[16] Also, 2,3,4-triiodobenzoic (TIBA) had a promotive effect on embryo formation.[17]

We also reported that calli derived from *Oncidium* roots, leaves, and stems all showed their embryogenic ability for plant regeneration via embryo formation,[18] and auxins and cytokinins played a role in embryogenesis similar to direct embryogenesis.

Such embryogenesis from a variety of tissue explants of *Oncidium* is easy to induce in large quantity. The usefulness of direct embryogenesis for genetic transformation was clearly demonstrated in a recent report on transformation attempt in *Oncidium*. Seed-derived PLBs were bombarded with *psSPFLP* containing genes encoding a sweet pepper ferredoxin-like protein (*pflp*), hygromycin phosphotransferase, and *b*-glucuronidase (GUS) driven by the cauliflower mosaicvirus (CaMV) 35S promoter. PCR analysis confirmed the success of the transformation (Li *et al.*[19]).

3.4 *Phalaenopsis*

Ten years ago, the use of flower stalks[20] and cytokinin-induced nodes[21] became popular for mass propagation of *Phalaenopsis*. In addition, PLB proliferation in liquid medium for mass propagation has been reported.[22] A paper on *Phalaenopsis* regeneration from callus culture via somatic embryogenesis has generated much attention.[23] The authors believed that their histological observations of the early young stage of PLBs regenerated from the primary callus are somatic embryos, with no variation in flowering plants regenerated through somatic embryogenesis. However, lacking were data and descriptions on the subculture of the primary callus culture. Chen *et al.*[24] reported a reliable protocol for plant regeneration from subcultured callus of *Phalaenopsis nebula*. The authors still used the term "PLB" for the early structure of the regeneration (Fig. 3.4).

Successive mass culture of PLBs in suspension culture for mass propagation is promising. *Doritaenopsis* callus cells can be subcultured in suspension culture, and the suspension cells can be cryopreserved.[25] Young *et al.*[26] demonstrated mass multiplication of PLBs in a bioreactor system and subsequent plant regeneration. Tokuhara and Mii[27] obtained embryogenic callus and cell suspensions from shoot tips of flower stalk buds and later found that carbohydrate sources played a crucial role in somatic embryogenesis in cell suspension cultures.[28]

The thin-cell layer method has been an efficient means for plant regeneration of orchids. A thin-section culture system for rapid regeneration

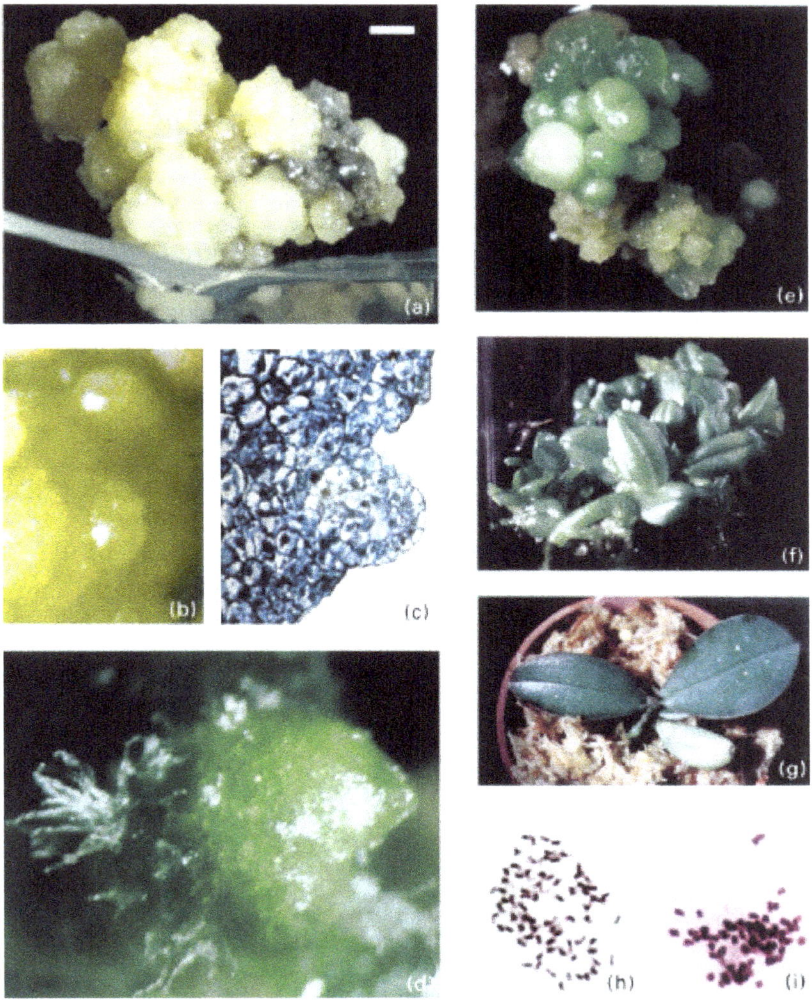

Fig. 3.4. Plant regeneration for callus culture of *Phalaenopsis* Nebula: protocorm-like body (PLB)-derived sub-culturable callus (a); embryo-like structures on the surface of callus (b); longitudinal section of callus showing a pro-embryo-like structure formed on the surface of the callus (c); PLBs formed on subcultured green callus (d); cluster of PLBs derived from green calli (e); regenerated plantlets *in vitro* (f); potted regenerants grown in sphagnum moss in the greenhouse (g); 76 chromosomes of a metaphase cell form a rot-tip squash of a regenerant (h); root-tip squash showing chromosomes of the donor plant.
(*From* Chen *et al.* (2000) *In Vitro/Plant* **46**:420–423.)

of the monopodal orchid hybrid *Aranda* has been developed by Lakshmanan *et al.*[29] Le *et al.*[30] obtained high-frequency shoot regeneration from *Rhynchostylis gigantean* using thin cell layers of leaves. Park *et al.*[31] developed an efficient and rapid method of PLB regeneration of *Doritaenopsis* that utilizes thin sections obtained from leaves as explants. Root tips also produced PLBs on a defined medium supplemented with TDZ.[32]

Direct somatic embryogenesis occurs in a culture of *Phalaenopsis*. A simple protocol for regenerating a *Phalaenopsis* cultivar through direct somatic embryogenesis on leaf explants in a defined medium plus TDZ (Fig. 3.5) was established.[33,34] Repetitive production of embryos involving secondary embryogenesis could be obtained by culturing segments of embryogenic mass on TDZ-containing medium, and plant conversion from embryos was successfully achieved on regulator-free growth medium. Basically, the plantlet obtained through direct somatic embryogenesis originates from a single cell origin. Therefore, the method is suitable for regenerating transgenic plants and could avoid the formation of chimera.

3.5 *Dendrobium*

Highlights among the *Dendrobium* literature can be categorized into three groups: shoot-bud proliferation, regeneration from callus, and direct somatic embryogenesis. Nayak[35] developed a TDZ-induced high-frequency shoot proliferation of two *Dendrobium* spp. High-frequency proliferation from thin cross-sections of PLBs of *Dendrobium nobile* was achieved in a defined medium.[36] Plant regeneration *of D. fimbriatum* from shoot-tip-derived callus has been documented.[37] Encapsulated PLBs has been attempted for long-term storage of the seed-derived PLBs.[38] Tricontanol (TRIA), a long 30-carbon primary alcohol, also a naturally occurring plant growth promoter, has been used for promoting shoot proliferation from cultured shoot tips of *D. nobile*.[39]

Several medicinally important orchids have been intensively studied in Taiwan over the last decade. Asymbiotic germination of immature seeds, plant development, and *ex vitro* establishment of plants of Shi-hu (*D. tosaense*) has been studied in detail by Tsay's group.[40] Nodal

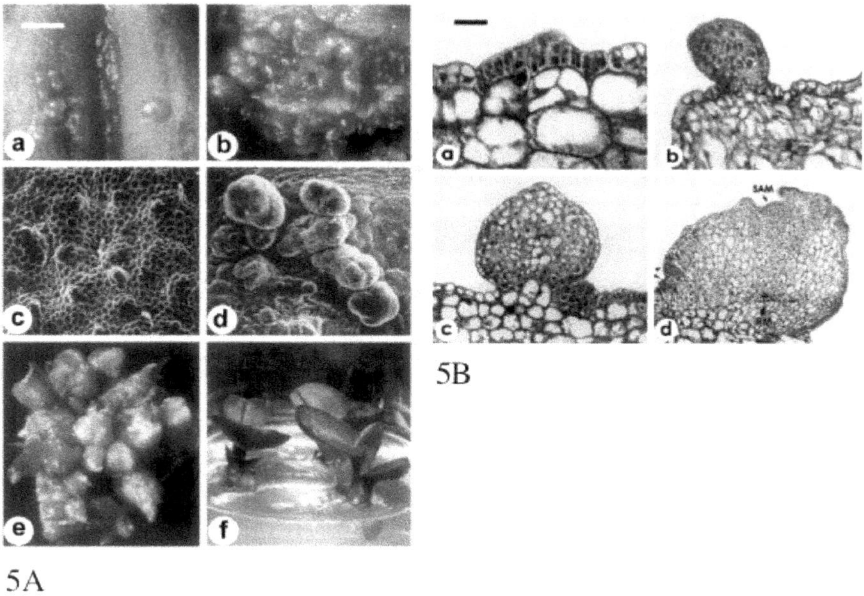

Fig. 3.5. Direct somatic embryogenesis for leaf explants of *Phalaenopsis* "Little Steve" and subsequent plant regeneration.

3.5A Somatic embryos formed directly from the leaf surface after 30 d of culture (a); embryos formed directly from the adaxial surface and the wounding of a leaf explant (b); Scanning electron microscopy (SEM) photograph of somatic pre-embryos formed from the leaf surface (c); SEM photograph of leaf-derived embryos (d); cluster of leaf-derived embryos forming shoots (e); regenerated plantlets (f).

3.5B Histology of direct somatic embryogenesis from leaf explants. Embryogenic cells originating from the leaf epidermal cell layer (a); a globular embryo contains densely stained and smaller embryonic cells than leaf cells (b); a developing embryo (c); a mature embryo, so-called protocorm, consisting of a shoot apical meristem and root meristem, and meristematic cells, formed from the basal region, were the origin of secondary embryogenesis (d).

(*From* Kuo *et al.* (2005). *In Vitro/Plant* **41**:453–456.)

segments have been used for mass propagation of *D. candidum*, a Chinese medicinal plant, through lateral bud proliferation.[41]

Direct organogenesis and direct somatic embryogenesis in *Dendrobium* spp. *in vitro* have been documented recently. Cytokinins were found causing direct somatic embryo formation on leaf explants of *Dendrobium chiengmai* Pink (Fig. 3.6).[42,43] Direct shoot formation from

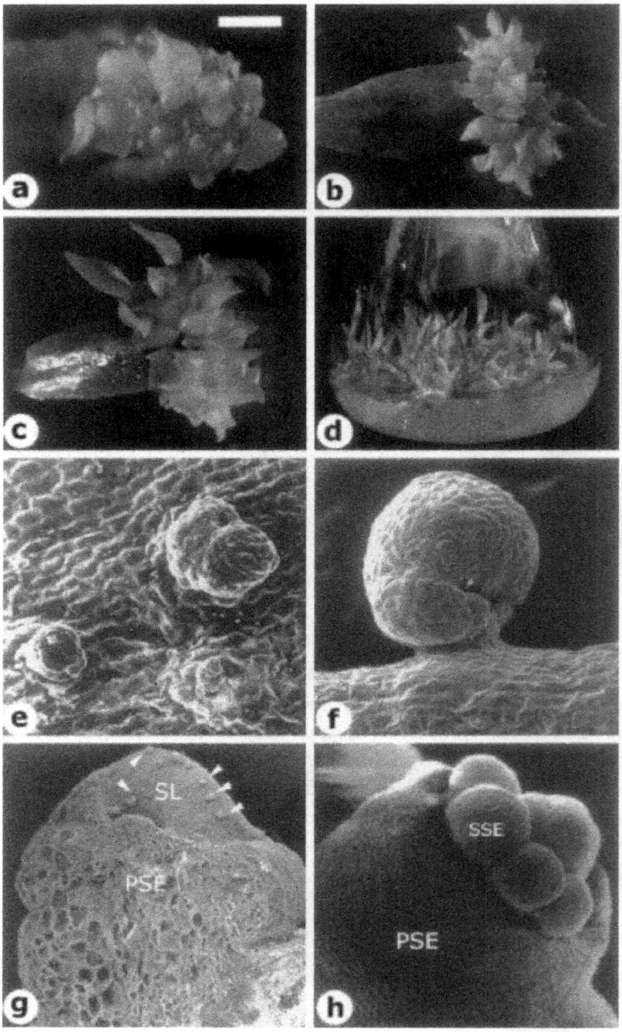

Fig. 3.6. Direct somatic embryogenesis from leaf explants of *Dendrobium* "Chiengmai Pink." Somatic embryos directly formed from leaf tip after 30 d of culture (a); embryo formed on the cut end (b); leaf-derived embryos formed shoots (c); regenerants (d); scanning electron microscopy (SEM) photograph of embryos were originating from leaf epidermal cells (e); an embryo in globular shape (f); secondary somatic embryos formed from the sheath leaf of a primary somatic embryos (g); secondary embryos in globular shape (h).
(*From* Chung *et al.* (2005). *In Vitro/Plant* **41**:765–769.)

foliar explants and PLBs has been developed for *in vitro* propagation of two *Dendrobium* hybrids.

Lin's group[44] demonstrated a successive transformation on a dendrobium to resist cymbidium mosaic virus.

3.6 *Paphiopedilum* and *Cypripedium*

Not much research has been documented in the past 10 years on lady's slippers. Efficient protocols for seed germination of slippers for propagation are still urgently needed. Regeneration from cultured explants is still considered difficult. Totipotent calli of a *Paphiopedilum* hybrid were induced from seed-derived protocorms on a defined medium containing 2,4-D and TDZ (Fig. 3.7).[45] Huang *et al.*[46] released a mericloning protocol enabling shoot multiplication and rooting of *Paphiopedilum* in a single medium. Nodal explants were useful for mass shoot bud proliferation on two *Paphiopedilum* hybrids (Fig. 3.7).[47] Direct multiple shoot bud proliferation for leaf explants of two *Paphiopedilum* hybrids clearly demonstrated that a practical mass propagation protocol can be developed from the morphogenesis.[48]

In the case of *Cypripedium*, five papers on improving seed germination have been published in the last decade. Cold temperature pre-treatment in seeds at 4°C is crucial for seed germination of *C. candidum*.[49] A fungus isolated from the roots of *C. macranthos* var *rebunense* induced symbiotic germination. Cold treatment of seeds at 4°C prior to fungal inoculation was required for symbiotic germination. Changing the timing of inoculation of the fungus to the seeds greatly improved germination frequency.[50] Totipotent callus of *C. formosanum*, an endangered slipper orchid species, was induced for seed-derived protocorm segments.[51] Protocorms were regenerated from the callus on a medium supplemented with BA only and, eventually, developed into plants. The age of seed capsule is crucial for good germination in *C. formosanum*.[52] Yan *et al.*[53] recently demonstrated an efficient way to mass-propagate *C. flavum* through multiple shoots of seedlings derived from mature seeds. The authors demonstrated that the efficiency in plantlets from the multiple shoots was much higher than that through PLBs derived from seeds.

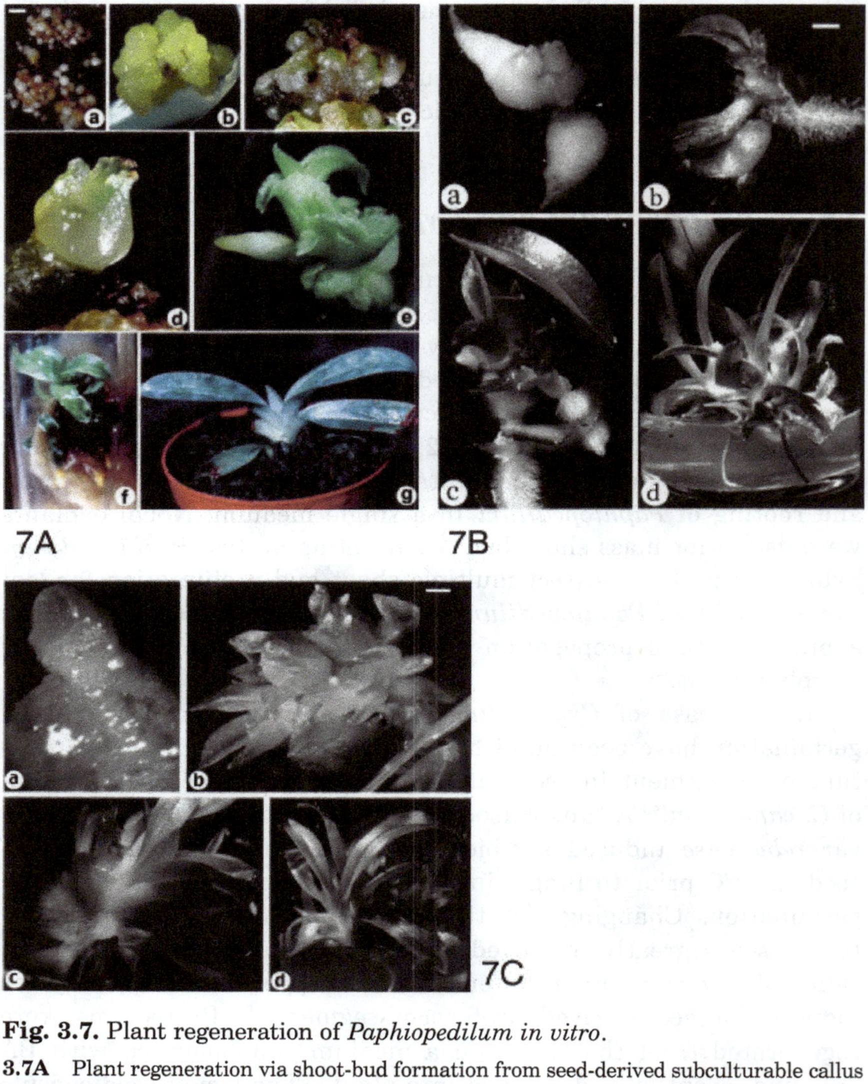

Fig. 3.7. Plant regeneration of *Paphiopedilum in vitro*.

3.7A Plant regeneration via shoot-bud formation from seed-derived subculturable callus (*From* Lin *et al.* (2000). *Plant Cell Tiss Org Cult* **62**:21–25.)

3.7B Plant regeneration via direct shoot-bud formation from nodal tissue of dwarf stem (Chen *et al.* (2002). *In Vitro/Plant* **38**:595–597.)

3.7C Plant regeneration through direct shoot-bud formation from leaf explants (*From* Chen *et al.* (2004). *Plant Cell Tiss Org Cult* **76**:11–15.)

3.7 *Epidendrum*

Sympodium orchid *Epidendrum radicans* was successively propagated via multiple-shoot proliferation from stem explants.[54]

3.8 Outlook: How Basic Research can Help the Orchid Industry

Any orchid biotechnology research should be relevant to the needs of orchid growers in any country and the benefits of the research transmitted to consumers at an affordable price. For the challenges ahead in orchid biotechnology and its practical application, I strongly encourage integrating modern technologies with classical breeding and physiological studies for the future success of the orchid industry.

The physiological basics of crop yield have been dealt with in great detail for most agricultural crops. Of course, orchids are no exception. In starting an orchid business or research, an important consideration is to ensure a steady supply of plant materials. Obtaining plant materials through conventional vegetative propagation methods is slow and costly. Today, the supply of uniform clonal planting materials comes, mainly from *in vitro* culture. The demand for micropropagated orchids also explains the recent, rapid increase in the number of commercial orchid laboratories and companies operating in Asia.

Clonal propagation of orchids, involving batch tissue culture, has been the mainstay throughout the world since 1960. Many problems are associated with the *in vitro* method and the batch tissue culture approach, as you can see in the presentations today. For example, batch culture is essentially a closed system, and *in vitro* conditions change with time and may not be optimal for cell growth. To optimize cell growth, all the factors must be maintained at optimal conditions. However, in batch culture, this condition is only possible by frequent subculturing. Subculturing involves considerable time and effort and will certainly increase production costs. Thus, improved cultural methodology is essentially based on a better understanding of basic physiology.

Furthermore, orchid seedlings grown in flasks are first transferred to a community pot, then to thumb pots, and then to a larger typical commercial pot. The duration of each transfer is about 3- to 6-months. Surprisingly, few scientific studies have investigated the growth and survival rate of plantlets during and after transfer from culture flasks to community pots in the greenhouse. In fact, high plantlet mortality rates have often been showed with some orchids. The hardening or acclimatization of plants in flasks and community pots certainly deserves more research.

More than two years are required for the orchid plantlets to reach the flowering stage. Orchids, particularly those with an epiphytic origin, are notoriously slow-growing plants. In their natural habitat, epiphytes usually meet with a greater degree of environmental stress, such as the supply of water and minerals. An understanding of how these orchids cope physiologically with environmental stress is important in order to improve their cultivation.

Flowering production is a major concern of any orchid industry. Flowering production depends on the genetic make-up of the orchid hybrids and how well they are grown. To achieve maximal flower yield, proper agronomic practices must be understood. Equally important is the control of flowering to meet market demand. The ability to control flowering in tropical orchids by physiological tools is indeed crucial.

The importance of proper post-harvest handling of cut-flowers has often been overlooked in the orchid cut-flower industry. The lack of proper post-harvest management in the industry is attributed to the paucity of information on the post-harvest physiology of orchid flowers.

New findings on the formation of direct embryos and shoot-buds for various tissue explants and *in vitro* flowering of regenerated embryos and plantlets on several orchids may address only a few aspects of the basic studies on orchids. The new knowledge may someday help to support the research I have mentioned previously.

In the Chinese literature, *Lan* (which means orchid in Chinese), is often personified as a man of virtue who strives for self-discipline, champions principles, and does not succumb to poverty and distress. Confucius wrote some 2500 years ago the following: "*Lan* that grows in deep forests never withholds its fragrances even when no one appreciates it." As a biotechnology researcher of orchids, I would like to see more insightful researches on such unique plants in the years to come.

References

1. Chang WC, Hsing YE. (1980) *In vitro* flowering of embryoids derived from mature root callus of ginseng (*Panax ginseng*). *Nature* **284:**341–342.
2. Chang WC, Hsing YE. (1980) Plant regeneration through somatic embryogenesis in root-derived callus of ginseng (*Panax ginseng* C. A. Meyer). *Theor Appl Genet* **57:**133–135.
3. Lin CS, Chen CT, Lin CC, Chang WC. (2003) A method for inflorescence proliferation. *Plant Cell Rep* **21:**838–843.
4. Lin CS, Lin CC, Chang WC. (2004) Effect of thidiazuron on vegetative tissue-derived somatic embryogenesis and flowering of *Bambusa edulis*. *Plant Cell Tiss Org Cult* **76:**75–82.
5. Chang C, Chang WC. (1998) Plant regeneration from callus culture of *Cymbidium ensifolium* var. misericors. *Plant Cell Rep* **17:**251–255.
6. Chang C, Chang WC. (2003) Cytokinins promotion of flowering in *Cymbidium ensifolium* var. misericors *in vitro*. *Plant Growth Reg* **39:**217–221.
7. Chang C, Chang WC. (2000) Effect of thidiazuron on bud development of *Cymbidium sinense* Willd *in vitro*. *Plant Growth Regul* **30:**171–175.
8. Huan LVT, Takamura T, Tanaka M. (2004) Callus formation and plant regeneration from callus through somatic embryo structures in *Cymbidium* orchid. *Plant Sci* **166:**1443–1449.
9. Kstenyuk I, So IS. (1999) Induction of early flowering in *Cymbidium niveo-narginatum* (Mark) *in vitro*. *Plant Cell Rep* **19:**1–5.
10. Chang C, Chen YC, Yen HF. (2005) Protocorm or rhizome? The morphology of seed germination in *Cymbidium dayanum* reichb. *Bot Bull Academia Sinica* **46:**71–74.
11. Chen JT, Chang C, Chang WC. (1999) Direct somatic embryogenesis on leaf explants on *Oncidium* Grower Ramsey and subsequent plant regeneration. *Plant Cell Rep* **19:**144–149.
12. Chen JT, Chang WC. (2000). Plant regeneration via embryo and shoot bude formation from flower-stalk explants of *Oncidium* Sweet Sugar. *Plant Cell Tiss Org Cult* **62:**95–100.
13. Chen JT, Chang WC. (2001) Effects of auxins and cytokinins on direct somatic embryogenesis on leaf explants of *Oncidium* Grower Ramsey. *Plant Growth Reg* **34:**229–232.
14. Chen JT, Chang WC. (2002) Effects of tissue culture conditions and explant characteristics on direct somatic embryogenesis on *Oncidium* Grower Ramsey. *Plant Cell Tiss Org Cult* **69:**41–44.
15. Chen JT, Chang WC. (2003) 1-aminocyclopropane-1-carboxylic ancid enhanced direct somatic embryogenesis from *Oncidium* leaf cultures. *Biol Plant* **46:**455–458.

16. Chen JT, Chang WC. (2003) Effects of GA3, ancymidol, cycocel and paclobutrazol on direct somatic embryogenesis of *Oncidium in vitro*. *Plant Cell Tiss Org Cult* **72:**105–108.
17. Chen JT, Chang WC. (2004) TIBA affects the induction of direct somatic embryogenesis from leaf explants of *Oncidium*. *Plant Cell Tiss Org Cult* **79:**315–320.
18. Chen JT, Chang WC. (2000) Efficient plant regeneration through somatic embryogenesis form callus cultures of *Oncidium* (Orchidaceae). *Plant Sci* **160:**87–93.
19. Li SH, Kuoh CS, Chen YH, *et al*. (2005) Osmotic sucrose enhancement of single-cell embryogenesis and transformation efficiency in *Oncidium*. *Plant Cell Tiss Org Cult* **81:**183–192.
20. Chen YQ, Piluek C. (1995) Effects of thidiazuron and N-6-beneyladaminopurine on shoot regeneration of *Phalaenopsis*. *Plant Growth Reg* **16:**99–101.
21. Duan DX, Chen H, Yazawa S. (1996) *In vitro* propagation of *Phalaenopsis* via culture of cytokinin-induced nodes. *J Plant Growth Reg* **15:**133–137.
22. Park YS, Kakuta S, Kano A, Okabe M. (1996) Efficient propagation of protocorm-like bodies of *Phalaenopsis* in liquid medium. *Plant Cell Tiss Org Cult* **45:**79–85.
23. Ishii Y, Takamura T, Goi M, Tanaka M. (1998) Callus induction and somatic embryogenesis of *Phalaenopsis*. *Plant Cell Rep* **17:**446–450.
24. Chen YC, Chang C, Chang WC. (2000) A reliable protocol for plant regeneration form callus culture of *Phalaenopsis*. *In Vitro/Plant* **36:**420–423.
25. Tukazaki H, Mii M, Tokuuhara K, Ishikawa K. (2000) Cryopreservation of diritaenopsis suspension culture by vitrification. *Plant Cell Rep* **19:**160–164.
26. Young PS, Murthy HN, Yoeup PK. (2000) Mass multiplication of protocorm-like bodies using bioreactor system and subsequent plant regeneration in *Phalaenopsis*. *Plant Cell Tiss Org Cult* **63:**67–72.
27. Tokuara K, Mii M. (2001) Induction of embryogenic callus and cell suspension culture from shoot tips excised from flower stalk buds of *Phalaenopsis* (Orchidaceae). *In Vitro/Plant* **37:**457–461.
28. Tokuara K, Mii M. (2003) Highly-efficient somatic embryogenesis from cell suspension cultures of *Phalaenopsis* orchids by adjusting carbohydrate sources. *In Vitro/Plant* **39:**635–639.
29. Lakshmanan P, Loh CS, Goh CJ. (1995) An *in vitro* method for rapid regeneration of manopodial orchid hybrid *Aranda* Deborah using thin section culture. *Plant Cell Rep* **14:**510–514.
30. Le BV, Phuong NTH, Hong LTA, *et al*. (1999) High frequency shoot regeneration from *Rhynchostylis gigantean* (Orchidaceae) using thin cell layers. *Plant Growth Reg* **28:**178–185.

31. Park SY, Yeung EC, Chakrabarty D, Paek KY. (2002) An efficient direct induction of protocorm-like bodies form leaf subepidermal cells of *Doritaenpsis* hybrid using thin-section culture. *Plant Cell Rep* **21:**46–51.
32. Park SY, Murthy HN, Peak KY. (2003) Protocorm-like body induction and subsequent plant regeneration from root tip cultures of *Doritaenopsis*. *Plant Sci* **164:**919–923.
33. Ku HL, Chen JT, Chang WC. (2005) Efficient plant regeneration through direct somatic embryogenesis from leaf explants of *Phalaenopsis* "little Steve." *In Vitro/Plant* **41:**453–456.
34. Chen JT, Chang WC. (2006) Direct somatic embryogenesis and plant regeneration from leaf explants of *Phalaenopsis anabilis*. *Biol Plant* **50:**169–173.
35. Nayak NR, Rath SP, Patnaik S. (1997) *In vitro* propagation of three epiphytic orchids *Cymbidium aloifolium* (L.) Sw., *Dendrobium apyllum* (Roxb.) fishch. and *Dendrobium moschatum* (Buch-Ham) Sw. Through thidiazuron-induced high frequency shoot proliferation. *Sci Hort* **71:**243–250.
36. Nayak NR, Sahoo S, Patnaik S, Rath SP. (2002) Establishment of thin cross section (TCS) culture method for rapid micropropagation of *Cymbidium aloifolium* (L.) Sw. and *Dendrobium nobile* Lindl. (Orchidaceae). *Sci Hort* **94:**107–116.
37. Roy J, Banerjee N. (2003) Induction of callus and plant regeneration form shoot-tip explants of *Dendrobium fimbriatum* Lindl. Var. oculatum HK.f. *Sci Hort* **97:**333–340.
38. Saiprasad GVS, Polisetty R. (2003) Propagation of three orchid genera using encapsulated protocorm-like bodies. *In Vitro/Plant* **39:**42–48.
39. Malabadi RB, Mulgund GS, Kallappa N. (2005) Micropropagation of *Dendribium nobile* form shot tip sections. *J Plant Physiol* **162:**473–478.
40. Lo SF, Nalawade SM, Kuo CL, *et al.* (2004) Asymbiotic germination of immature seeds, plantlet development and *ex vitro* establishment of plants of *Dendrobium tosaense* makino — A medicinally important orchid. *In Vitro/Plant* **40:**528–535.
41. Shiau YH, Malawade SM, Hsia CN, *et al.* (2005) *In vitro* propagation of the Chinese medicinal plant, *Dendrobium candidum* Wall, ex Lindl., from axenic nodal segments. *In Vitro/Plant* **41:**666–670.
42. Chung HH, Chen JT, Chang WC. (2005) Cytokinins induce direct somatic embryogenesis of *Dendrobiu chiengmai* Pink and subsequent plant regeneration. *In Vitro/Plant* **45:**765–769.
43. Chung HH, Chen JT, Chang WC. Cytokinins induce direct somatic embryogenesis from leaf explants of *Dendrobiu chiengmai* Pink. *Biol Plant* (in press).
44. Chang C, Chen YC, Hsu YH, *et al.* (2005) Transgenic resistance to *Cymbidium mosaic virus* in *Dendrobium* expressing the viral capsid protein gene. *Transgenic Res* **14:**41–46.

45. Li YH, Chang C, Chang WC. (2000) Regeneration from callus culture of *Paphiopedilum* hybrid. *Plant Cell Tiss Org Cult* **62:**21–25.
46. Huang LC, Lin CJ, Kuo CI, *et al.* (2001) *Paphiopedilum* cloning *in vitro*. *Sci Horti* **91:**111–121.
47. Chen TY, Chen JT, Chang WC. (2002) Multiple shoot formation and plant regeneration from stem nodal explants of *Paphiopedilum* orchids. *In Vitro/Plant* **38:**595–597.
48. Chen TY, Chen JT, Chang WC. (2004) Plant regeneration through direct shoot bud formation from leaf cultures of *Paphiopedilum* orchids. *Plant Cell Tiss Org Cult* **76:**11–15.
49. Shimura H, Koda Y. (2004) Micropropagation of *Cypripedium macranthos* var. rebunense through protocorm-like bodies derived from mature seeds. *Plant Cell Tiss Org Cult* **78:**273–276.
50. Shimura H, Koda Y. (2005) Enhanced symbiotic seed germination of *Cypripedium macranthos* var. rebunense following inoculation after cold treatment. *Physiol Plant* **123:**281–287.
51. Lee YI, Lee N. (2003) Plant regeneration from protocorm-derived callus of *Cypripedium formosanum*. *In Vitro/Plant* **39:**475–479.
52. Lee YI, Lee N, Yeung EC, Chung MC. (2005) Embryo development of *Cypripedium formosanum* in relation to seed germination *in vitro*. *J Amer Soc Horti Sci* **130:**747–753.
53. Yan N, Hu H, Huang Jl, *et al.* (2006) Micropropagation of *Cypripedium flavum* through multiple shoots of seedlings derived from mature seeds. *Plant Cell Tiss Org Cult* **84:**113–117.
54. Chen LR, Chen JT, Chang WC. (2002) Efficient production of protocorm-like bodies and plant regeneration from flower stalk explants of the sympodial orchid *Epidendrum Radicans*. *In Vitro/Plant* **38:**441–445.

Chapter 4

Somaclonal Variation in Orchids

Fure-Chyi Chen[*,†] *and Wen-Huei Chen*[‡]

Mass production of orchids is achieved through micropropagation of axillary buds of flower stalks or shoot meristem culture. Somaclonal variation occurs during proliferation of both shoots and protocorm-like bodies (PLBs) and leads to morphological or physiological changes in the finished potted plants. We observed both vegetative and reproductive variants, depending on the cultivar or genetic background, among tissue-cultured orchid plants. The use of molecular approaches such as random amplified polymorphic DNA (RAPD), cDNA-RAPD and cDNA suppression subtractive hybridization has resulted in the successful identification of expressed sequence tags (ESTs) from wild-type and peloric flower buds of several *Phalaenopsis* and *Doritaenopsis* hybrids. Several candidate genes were cloned and transcript levels compared in these plants. The results suggested that some genes, such as a retroelement, could have been abnormally activated in the peloric mutants. We discuss other investigations, such as DNA methylation status, which might also play a role in somaclonal variations in orchids.

4.1 Introduction

Somaclonal variation refers to the variation seen in plants regenerated from tissue culture of any plant tissue. Earlier work on

*Corresponding author.

†*Department of Plant Industry and Graduate Institute of Biotechnology, National Pingtung University of Science & Technology, Pingtung, Taiwan.*

‡*Department of Life Sciences, National University of Kaohsiung, Kaohsiung, Taiwan.*

somaclones aimed at selecting for breeding valuable genotypes such as variant protoclones inheriting new phenotypes[1,2] or with altered ploidy levels.[3,4] For other crops, interest was in selecting somaclonal variants with disease resistance, low-temperature tolerance and other traits;[5–7] (for detailed reviews of somaclonal variation, see Refs. 1, 7, 8).

Orchids are grown worldwide for their commercial value.[9] In Taiwan, *Phalaenopsis*, *Oncidium* and several other orchid genera have been developed quickly in response to market demand. A novel orchid variety selected from a seedling population results from cross-hybridization of two parents with desirable traits. The selected plant is then propagated from either axillary buds of flower stalks or shoot meristem culture *in vitro*. Such mass propagation can result in several hundred thousands of plantlets with uniform growth and flowering time.

Some micropropagated plantlets of certain hybrids may appear abnormal during the vegetative or reproductive stages because of somaclonal variation, which is a concern for orchid growers because the variants may affect pot plant quality. The mechanism of these variations during tissue culture is still largely unknown. Regarding somaclonal variation of other crop plants, the past two decades have seen investigations into DNA methylation status, activation of transposable elements or retrotransposons, and histone modifications.[10–16]

This chapter covers the methods of tissue culture, somaclonal variation, and molecular approaches to understanding the mechanisms of somaclonal epigenetic changes in exploring novel strategies for lowering the somaclonal variation rate during tissue culture of orchids.

4.2 Orchid Tissue Culture and Plantlet Production

To evaluate horticultural traits or performance of a hybrid of two desirable cultivars, plants of a small population of the same clones derived from tissue culture must be grown in potting medium. For tissue culture, first an elite orchid plant from the hybrid combination is selected, usually during the flowering stages, then an axillary bud is excised from a pseudobulb axil or flower-stalk node for use as starting explant materials. The materials are surface sterilized with diluted sodium hypochloride containing surfactant and washed with sterile water several times. The explants are then cultured on a suitable nutrient medium, such as Hyponex-based medium supplemented with cytokinin and auxin,[17] and plantlets are potted.

Oncidium and *Phalaenopsis* orchids have been successfully propagated by culture of flower-stalk nodes.[17,18] Generally, two methods of micropropagation are adopted by the industry, one through multiplication of adventitious buds and the other through the induction and subsequent proliferation of protocorm-like bodies (PLBs). PLBs have been successfully induced from *in vitro*-grown etiolated leaves of *Phalaenopsis aphrodite*.[19]

4.3 Somaclonal Variation in Orchids

Orchid mutants can be derived from variations during tissue culture or from seedling mutations. Since these days mericloned plants are preferred for the flower market,[20] one can look for orchid plants with abnormal morphology during visits to orchid nurseries. We grew such plants in the greenhouse for at least two flowering seasons to further observe their stability. From our observations, the mutations could be divided into two categories: those affecting vegetative growth and those rendering reproductive abnormalities, such as peloric or semipeloric flowers on which the lateral petals are completely or partially converted into labellum (lip)-like structures.[21]

We obtained several vegetative mutations, including overabundant leaves, multiple axillary shoots, distorted sickle-shaped leaves, and polyploidy (Fig. 4.1). In reproductive mutants, peloric and semipeloric flowers have been observed in mericloned plants of many *Phalaenopsis* and/or *Doritaenopsis* hybrids, with semipeloric flowers being more common (Fig. 4.2). Since *Doritaenopsis* hybrids are derived from cross-hybridization of *Phalaenopsis* and *Doritis* genotypes, their progenies tend to be more unstable genetically than those of *Phalaenopsis* hybrids because of variation in chromosome configuration or genome size.[21,22] Mutations affecting internode length were reported in tissue culture-derived plants of *Phalaenopsis* and *Doritaenopsis* orchids.[23]

Peloric flower mutations vary depending on the degree of severe conversion of petals into lip-like protrusions or narrowed petals. In some

Fig. 4.1. Vegetative mutants in orchids. (**A**) Carotenoid accumulation in leaves and inflorescence of a *Phalaenopsis aphrodite*; (**B**) Sickle-shaped leaves in a *P.* Hwafeng Redjewel mutant plant; (**C**), A *Doritaenopsis* Purple Gem mutant with an overabundance of leaves; (**D**) A *Dtps*. mutant with flower-like structures on the leaves; (**E**), A hybrid of *Dtps*. Sogo Pride mutant with multiple axillary shoots; (**F**) Normal plants of *Phalaenopsis* mericlones.

Fig. 4.2. Reproductive mutants in orchids. (**A**) Mutation with petaloid column; (**B**) A semi-peloric flower with labellum-like lateral petals; (**C**) Sepal mutation with protrusion on the upper epidermal tissues of lateral sepals; (**D**) Mutation possessing anther-like structures in the distal part of lateral petals; (**E**) Yellow peloric flowers; (**F**) Anthocyanin pattern mutation in a *Doritaenopsis* hybrid, wild-type on the left and color-stripe mutant on the right.

severe cases, the lateral petals are changed completely into lip structures, and the plants lack pollinia.

Another class of mutations involves serrated petals and labellum. The remaining plant tissues are not affected. The serrated petals are visible only during the flowering stages. The cause of this mutation is not clear. A serrated-petal mutation in *Arabidopsis* was caused by altered sterol composition and ecotopic endoreduplication in petal tips.[24] Methods successful in *Arabidopsis* study could be used to investigate whether endoreduplication caused the serrated petals of our *Phalaenopsis* mutants and to clone the candidate gene for sterol biosynthesis.

4.4 Strategy for Studying Somaclonal Variation

Since the generation time for orchids is relatively long, conventional cross-hybridization or selfing to learn the inheritance of somaclonal variation is difficult. Unconventional molecular approaches to study epigenetic mutations in other organisms have involved suppression subtractive hybridization (SSH);[25,26] differential display;[27,28] molecular markers such as random amplified polymorphic DNA (RAPD);[29–33] amplified fragment-length polymorphism (AFLP);[34,35] methylation-sensitive restriction FLP (RFLP);[13] and candidate gene cloning by use of degenerate primers and PCR based on conserved amino acid sequences of target genes. Flow cytometry has been used to study endopolyploidy of embryos or PLBs in *Vanda* and *Doritaenopsis* orchids during *in vitro* germination, with ploidy levels enhanced by treatment with the auxins N-acetyl aspartate or 2,4-dichlorophenoxyacetic acid.[36,37]

4.4.1 *Molecular markers*

Somaclones derived from *in vitro* culture of flower-stalk nodes of the *Phalaenopsis* cultivar True Lady were studied by use of RAPD markers and isozyme electrophoretic patterns. With the use of 38 selected random primers for PCR amplification of genomic DNAs, polymorphism was revealed in somaclones with deformed flower shapes,[30] and with two isozymes, viz., aspartate aminotransferase and phosphoglucomutase, wild types were differentiated from variants.[30] cDNA-RAPD analysis revealed differences in mRNA expression in both the wild type and peloric and semipeloric variants in *P.* Little Mary.[26]

4.4.2 *cDNA suppression subtractive hybridization*

Differentially expressed genes have been effectively studied by use of cDNA SSH[25] and modified SSH methods such as SSH followed by negative subtraction chain (NSC.)[38] Both SSH or NSC techniques can enrich target sequences at different efficiency rates, but background signals may be a concern. We adopted SSH to compare transcripts enriched in young flower buds of both the wild-type and peloric *Phalaenopsis* and *Doritaenopsis* hybrids.[26] One problem encountered with SSH of *Phalaenopsis* orchids is that often the plants were infected by viruses such as cymbidium mosaic virus (CymMV) and odontoglossum ringspot virus (ORSV), which may interfere with the effectiveness of the SSH approach. In one case, approximately 28% of the ESTs from the peloric mutants were due to CymMV sequences.[26] A similar result has been reported for *P.* Hsiang Fei with use of cDNA-AFLP.[39]

To confirm the expression level of selected ESTs resulting from our SSH analysis, we used real-time RT-PCR to compare the transcripts in both the wild-type and peloric and semipeloric flower buds. CymMV RNA-dependent RNA polymerase, a retroelement, and an auxin-regulated dual-specificity cytosolic kinase were preferentially upregulated in the peloric flower buds (Fig. 4.3) and to a lesser extent in semipeloric buds.[26] A TCP protein, reported to be responsible for the symmetric development of plant organs,[40] was slightly up-regulated in peloric flower buds.[26] Obtaining virus-free orchid plants will help in better adopting SSH for future analysis of transcript profiling in the somaclonal mutants.

4.4.3 *Methylation-specific restriction PCR assays*

DNA hypomethylation, which may lead to abnormal activation of certain genes, has been related to epigenetic changes in diseased animal tissues.[41] Methylation-sensitive restriction enzymes and quantitative PCR have been widely used to investigate methylation status in the promoter regions of target genes.[42,43] Mutation in histone H3 methylation may also contribute to DNA methylation changes.[44,45] McrBC, a restriction enzyme that preferentially digests tracts of DNA flanked by methylcytosine residues, has been used to reveal genome-wide reduction in DNA methylation in the mantled phenotypes of regenerated oil palm.[16] The use of these techniques in investigating somaclonal variation in orchids may be profitable.

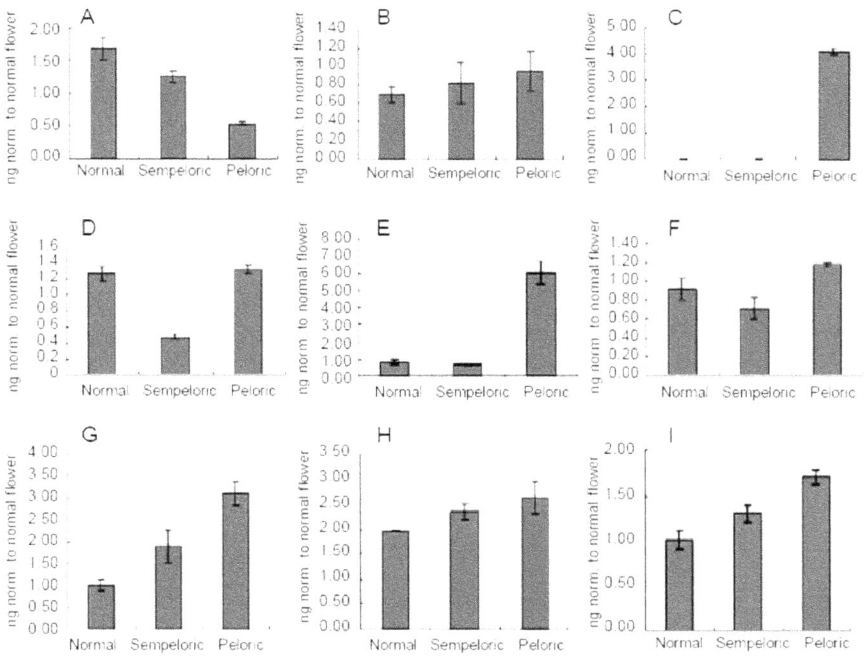

Fig. 4.3. Transcript profiles in wild-type, peloric and semi-peloric flower buds of *Phalaenopsis* Little Mary by real-time RT-PCR analyses. (**A**) a TGA1a-like protein; (**B**) a protein kinase; (**C**) CymMV RNA-dependent RNA polymerase; (**D**) a proline iminopeptidase; (**E**) a retroelement; (**F**) an unknown protein; (**G**) auxin-regulated dual specificity cytosolic kinase; (**H**) cyclophilin-like protein; (**I**) a transcriptional factor PCF6 or TCP family protein. (Chen *et al.*, 2005; Courtesy of Cell Research.)

4.4.4 *Cloning of candidate genes*

Peloric flowers show a change from bilateral to radial symmetry; however, an attempt to clone the TCP homolog of cycloidea,[46] based on conserved amino acid sequences by RT-PCR, was fruitless (Chen FC, unpublished result). Thus, the conserved amino acid sequences of *Phalaenopsis* orchids may differ from that of other higher plants. A candidate gene of the *knotted1*-like homeobox gene family,[47] which may have roles in the patterning of plant body plans, has been cloned by the design of degenerate primers following alignment of several protein sequences. The *knotted1*-like homeobox genes may play a role in the formation of the lobe in the labellum of *Phalaenopsis* flowers. Four B-class *Phalaenopsis DEF*-like MADS-box genes were detected in floral organs; the expression of *PeMADS4* was in the lips and column of the wild type, and it extended to the lip-like petals in the peloric mutant.[48]

4.5 Conclusions

We have obtained many somaclonal variants of *Phalaenopsis* and *Oncidium* orchids. Molecular and morphological studies have revealed abnormalities in the vegetative and reproductive tissues. Orchid viruses exist widely in orchid plants and may hinder some molecular analyses. Several candidate genes were preferentially expressed in the peloric mutants. The wide spectrum of somaclonal variation may contribute to our further understanding of the molecular basis of epigenetic changes at the DNA or chromatin levels. Recently, we cloned DNA methyltransferase from *Phalaenopsis*. Our preliminary study indicated its down-regulation in peloric mutants. Further characterization of this gene should offer an opportunity to elucidate its function in the DNA methylation status of orchid genomes.

4.6 Perspectives

Because epigenetic changes often occur in orchid tissue culture, the question is whether the phenomenon actually disturbs the developmental pattern of an orchid flower. Several approaches to investigation are available and await confirmation. Genetic transformation of candidate genes to reveal their function in orchids requires efficient tissue culture and regeneration, such as induction of PLBs from cultures of etiolated leaf segments.[19] The candidate genes can also be used to transform corresponding mutants of *Arabidposis* or rice to complement their defect. One can also use RNA interference to block the candidate genes *in planta* to abolish their morphological competency.

Acknowledgments

The work described in this chapter was funded by the Council of Agriculture, Taiwan, awarded to the first author. We thank the graduate students and research assistants in both authors' labs for assistance and support.

References

1. Bidney DL, Shepard JF. (2004) Phenotypic variation in plants regenerated from protoplasts: the potato system. *Biotechnol Bioeng* **23:**2691–2701.
2. Ramulu KS, Dijkhuis P, Roest S. (1984) Genetic instability in protoclones of potato (*Solanum tuberosum* L. cv. "Bintje"): new types of variation after vegetative propagation. *Theor Appl Genet* **68:**515–519.
3. Nelson RS, Karp A, Bright SWJ. (1986) Ploidy variation in *Solanum brevidens* plants regenerated from protoplasts using an improved culture system. *J Exp Bot* **37:**253–261.
4. Ramulu KS, Dijkhuis P, Roest S. (1983) Phenotypic variation and ploidy level of plants regenerated from protoplasts of tetraploid potato (*Solanum tuberosum* L. cv. "Bintje"). *Theor Appl Genet* **65:**329–338.
5. Heath-Pagliuso S, Pullman J, Rappaport L. (1988) Somaclonal variation in celery: screening for resistance to *Fusarium oxysporum* f. sp. *apii*. *Theor Appl Genet* 75:446–451.
6. Huitema JBM, Preil W, Gussenhoven GC, Schneidereit M. (1989) Methods for the selection of low-temperature tolerant mutants of *Chrysanthemum morifolium* Ramat. by using irradiated cell suspension cultures. I. Selection of regenerants *in vivo* under suboptimal temperature conditions. *Plant Breed* **102:**140–147.
7. Maliga P. (1984) Isolation and characterization of mutants in plant cell culture. *Ann Rev Plant Physiol* **35:**519–542.
8. Mohan Jain S. (2001) Tissue culture-derived variation in crop improvement. *Euphytica* **118:**153–166.
9. Griesbach RJ. (2002) Development of *Phalaenopsis* orchids for the mass-market. In: Janick J, Whipkey A (ed.), *Trends in New Crops and New Uses*. ASHS Press, Alexandria, VA, pp. 458–465.
10. Cheng C, Daigen M, Hirochika H. (2006) Epigenetic regulation of the rice retrotransposon Tos17. *Mol Genet Genomics* **276:**378–390.
11. Hirochika H, Sugimoto K, Otsuki Y, *et al*. (1996) Retrotransposons of rice involved in mutations induced by tissue culture. *Proc Natl Acad Sci USA* **93:**7783–7788.
12. Jaligot E, Rival A, Beule T, *et al*. (2000) Somaclonal variation in oil palm (*Elaeis guineensis* Jacq.): the DNA methylation hypothesis. *Plant Cell Rep* **7:**684–690.
13. Jaligot E, Beule T, Rival A. (2002) Methylation-sensitive RFLPs: Characterisation of two oil palm markers showing somaclonal variation-associated polymorphism. *Theor Appl Genet* **104:**1263–1269.
14. Jaligot E, Beule T, Baurens FC, *et al*. (2004) Search for methylation-sensitive amplification polymorphisms associated with the mantled variant phenotype in oil palm (*Elaeis guineensis* Jacq). *Genome* **47:**224–228.

15. Koukalova B, Fojtova M, Lim KY, *et al*. (2005) Dedifferentiation of tobacco cells is associated with ribosomal RNA gene hypomethylation, increased transcription, and chromatin alterations. *Plant Physiol* **139:**275–286.
16. Kubis SE, Castilho AM, Vershinin AV, Heslop-Harrison JS. (2003) Retroelements, transposons and methylation status in the genome of oil palm (*Elaeis guineensis*) and the relationship to somaclonal variation. *Plant Mol Biol* **52:**69–79.
17. Chen FC, Hsu CC. (2002) Techniques of orchid tissue culture. In: *Proceedings of the Current Situation of Floricultural Industry and Future Perspectives* Taiwan Floricultural Association, Taipei, pp. 26–36.
18. Chen FC, Chen TC. (1998) Effect of salt strength and organic additives on the *in vitro* growth of protocorm-like-bodies and plantlets of *Oncidium* Grower Ramsey. *J Chin Soc Hortic Sci* **44:**403–412.
19. Hsu CC, Chen FC. (2003) Plant regeneration from protocorm-like bodies induced in etiolated leaves of *Phalaenopsis aphrodite* Rchb. f. *J Chinese Soc Hort Sci* **49:**335–342.
20. Shen T-M, Hsu ST, Chen JJ, Chen FC. (2006) Research and development of *Phalaenopsis* industry in Taiwan. In: *Proceedings of the 2006 Taiwan Intl Orchid Symp*, Taiwan Orchid Growers Association, Tainan, Taiwan, pp. 7–13.
21. Chen FC, Chen YH, Lee WL, *et al*. (2004) Orchid mutants from tissue culture — implication of gene expression to orchid development. In: *Proceedings of the 8th Asia Pacific Orchid Conference*, Taiwan Orchid Growers Association, Tainan, Taiwan, pp. 314–328.
22. Lin S, Lee HC, Chen WH, *et al*. (2001) Nuclear DNA contents of *Phalaenopsis* sp. and *Doritis pulcherrima*. *J Amer Soc Hort Sci* **126:**195–199.
23. Tokuhara K, Mii M. (1998) Somaclonal variations in flower and inflorescence axis in micropropagated plants through flower stalk bud culture of *Phalaenopsis* and *Doritaenopsis*. *Plant Biotech (Japan)* **15:**23–28.
24. Hase Y, Fujioka S, Yoshida S, *et al*. (2005) Ectopic endoreduplication caused by sterol alteration results in serrated petals in *Arabidopsis*. *J Exp Bot* **56:**1263–1268.
25. Diatchenko L, Lau YF, Campbell AP, Chenchik A. (1996) Suppression subtractive hybridization: a method for generating differentially regulated or tissue-specific cDNA probes and libraries. *Proc Natl Acad Sci USA* **93:**6025–6030.
26. Chen, YH, YJ Tsai, JZ Huang, FC Chen. (2005) Transcription analysis of peloric mutants of *Phalaenopsis* orchids derived from tissue culture. *Cell Res* **15:**639–657.
27. Liang P, Pardee AB. (1992) Differential display of eukaryotic messenger RNA by means of the polymerase chain reaction. *Science* **257:**967–971.
28. Tregear JW, Morcillo F, Richaud F, *et al*. (2002) Characterization of a defensin gene expressed in oil palm inflorescences: induction during tissue

culture and possible association with epigenetic somaclonal variation events. *J Exp Bot* **53:**1387–1396.
29. Purwantoro A, Supaibulwatana K, Mii M, Koba T. (1999) Cytological and RAPD (random amplified polymorphic DNA) analysis of somaclonal variation in Easter lily (*Lilium longiflorum* Thinb). *Plant Biotechnol (Jpn)* **16:**247–250.
30. Chen WH, Chen TM, Fu YM, *et al.* (1998) Studies on somaclonal variation in *Phalaenopsis*. *Plant Cell Rep* **18:**7–13.
31. Martins M, Sarmento D, Oliveira MM. (2004) Genetic stability of micropropagated almond plantlets, as assessed by RAPD and ISSR markers. *Plant Cell Rep* **23:**492–496.
32. Palombi MA, Lombardo B, Caboni E. (2007) *In vitro* regeneration of wild pear (*Pyrus pyraster* Burgsd) clones tolerant to Fe-chlorosis and somaclonal variation analysis by RAPD markers. *Plant Cell Rep* **26**:489–496. [Epub ahead of print] 10.1007/s00299-006-0256-9
33. Rival A, Bertrand L, Beulé T, *et al.* (1998) Suitability of RAPD analysis for the detection of somaclonal variants in oil palm (*Elaeis guineensis* Jacq.). *Plant Breed* **117:**73–76.
34. Chen J, Henny RJ, Devanand PS, Chao CT. (2006) AFLP analysis of nephthytis (*Syngonium podophyllum* Schott) selected from somaclonal variants. *Plant Cell Rep* **24:**743–749.
35. Matthes M, Singh R, Cheah S-C, Karp A. (2001) Variation in oil palm (*Elaeis guineensis* Jacq.) tissue culture-derived regenerants revealed by AFLPs with methylation-sensitive enzymes. *Theor Appl Genet* **102:**971–979.
36. Lim WL, Loh CS. (2003) Endopolyploidy in *Vanda* Miss Joaquim (Orchidaceae). *New Phytol* **159**:279–287.
37. Mishiba K-I, Okamoto T, Mii M. (2001) Increasing ploidy level in cell suspension cultures of *Doritaenopsis* by exogenous application of 2,4-dichlorophenoxyacetic acid. *Physiol Plant* **112:**142–148.
38. Li L, Techel D, Gretz N, Hildebrandt A. (2005) A novel transcriptome subtraction method for the detection of differentially expressed genes in highly complex eukaryotes. *Nucleic Acids Res* **33:**e136.
39. Chen H-H, Tsai W-C, Lin S, Chen W-H. (2001) cDNA-AFLP analysis of differential gene expression in the flower buds of *Phalaenopsis* HSING FEI cv. H. F. and its somaclonal variant. *Plant and Animal Genome IX Conference*, p. 323.
40. Cubas P, Lauter N, Doebley J, Coen E. (1999b) The TCP domain: a motif found in proteins regulating plant growth and development. *Plant J* **18:**215–222.
41. Ortmann CA, Burchert A, Hölzle K, *et al.* (2005) Down-regulation of interferon regulatory factor 4 gene expression in leukemic cells due to hypermethylation of CpG motifs in the promoter region. *Nucleic Acids Res* **33:**6895–6905.

42. Germann S, Juul-Jensen T, Letarnec B, Gaudin V. (2006) DamID, a new tool for studying plant chromatin profiling *in vivo*, and its use to identify putative LHP1 target loci. *Plant J* **48:**153–163.
43. Ott T, Nelsen-Salz B, Doring HP. (1992) PCR-aided genomic sequencing of 5' subterminal sequences of the maize transposable element Activator (Ac) in transgenic tobacco plants. *Plant J* **2:**705–711.
44. Jackson JP, Lindroth AM, Cao X, Jacobsen SE. (2002) Control of CpNpG DNA methylation by the KRYPTONITE histone H3 methyltransferase. *Nature* **416:**556–560.
45. Johnson L, Cao X, Jacobsen S. (2002) Interplay between two epigenetic marks. DNA methylation and histone H3 lysine 9 methylation. *Curr Biol* **12:**1360–1367.
46. Cubas P, Vincent C, Coen E. (1999a) An epigenetic mutation responsible for natural variation in floral symmetry. *Nature* **401:**157–161.
47. Serikawa KA, Martinez-Laborda A, Zambryski P. (1996) Three knotted1-like homeobox genes in *Arabidopsis*. *Plant Mol Biol* **32:**673–683.
48. Tsai WH, Kuoh CS, Chuang MH, *et al.* (2004) Four DEF-like MADS box genes displayed distinct floral morphogenic roles in Phalaenopsis orchid. *Plant Cell Physiol* **45:**831–844.

Chapter 5

The Screening of Orchid Mycorrhizal Fungi (OMF) and their Applications

*Doris C. N. Chang**

Wild-grown and greenhouse-cultivated orchid roots and the associated fungi were isolated on 1/6 PDA. The fungi isolated from various Taiwan orchid roots were *Acremonium* spp.; *Alternaria* spp.; *Cylindrocarpon* spp.; *Fusarium* spp.; *mycelia sterile*; *Penecillum* spp.; *Rhizoctonia* spp.; and *Trichoderma* spp. Those isolates that showed no pathogenecity to the major crops of Taiwan: mungbean, cucumber, radish and rice seedlings were used for inoculation of orchid mycorrhizal fungi. Only those proving beneficial in the growth of orchids were considered as orchid mycorrhizal fungi (OMF). *Trichoderma* and *Rhizoctonia* were the most frequent fungi isolates from the Taiwan wild-grown orchid roots. Based on 10 years of inoculation tests, R01, R02, and R04 are shown to be the three isolates with OMF of commercial value. Among them, R01 and R02 are binucleate, while R04 is a multinucleate *Rhizoctonia* sp. Up till now the identification of R01 and R02 has been uncertain. However, it is known that R04 is *Rhizoctonia solani*, in the AG6, which has no pathogenic effects on orchids. Seed germinations were enhanced by R02 for *Anoectochilus formosanus* Hayata and R01 for *Haemaria discolor* var. *dawsoniana*. Growth of seedlings was also highly enhanced by the inoculation of *Rhizoctonia* OMF isolates for three medicinal uses of orchids, such as *A. formosanus* (R02, R04), *H. discolor* (R01), *and Dendrobium* spp. Higher acid and alkaline phosphatases and superoxide dismutase activities, together with

**Department of Horticulture, National Taiwan University, Taipei, Taiwan.*

flavonoids, phenolic compounds, polysaccharides, ascorbic acids, and phosphate contents were all higher in mycorrhizal tissues than the non-mycorrhizal control of *A. formosanus*. Vegetative and reproductive growth was stimulated by OMF for ornamental orchids, such as *Doritaenopsis* spp. (R02, R01, R04), *Phalaenopsis* spp. (R01, R02, R04), and *Phaphiopedium delenatii* (R02, R04). Oat meal agar should replace MS or Kyoto agar medium if OMF is inoculated *in vitro*. The physiological effects of OMF and their potential applications on commercially cultivated orchids are presented. It was highly recommended that the OMF be applied as follows: 1) to increase the survival rates of micropropagated plantlets or seedlings *ex vitro*; 2) to enhance both vegetative and reproductive growth of orchid plants; 3) to result in earlier flowering and enhanced flower quality; and 4) to reduce disease infection rates.

5.1 Introductions

It is well-established that orchids have fungi growing in their roots. Cells of the root cortex are occupied, to a greater or lesser extent, by clusters of fungal hyphae, and these occur under both natural and horticultural conditions. The orchid-fungus relationship is one of many known forms of mycorrhiza (the term means fungus-root), a common form of symbiosis between fungi and plants.[1] Many orchid mycorrhizal fungi belong to the genera *Rhizoctonia, Epulorrhiza, or Ceratorhiza.*[2]

Bernard[3] and Burgeff[4–7] studied seed germination of orchids and recognized that, in general, and particularly under natural conditions, successful germination was virtually impossible in the absence of a fungus. This implies that germination is not successful unless it is succeeded by growth. In the absence of infection, the growth rate is often negligible regardless of the availability of nutrients. Consequently, a symbiotic method for seed germination was developed, which involved culture tubes inoculated with both seeds and fungi.[7–13] This procedure was difficult for the commercial grower. It was also characterized by a low success rate.[1,3] Hadley[1] wrote a chapter called "Orchid mycorrhiza" in the book *Orchid Biology Reviews and Perspectives, II*,[14] an excellent volume on orchid mycorrhizal researches from 1960- to the 1980s. In contrast to symbiotic germination, the non-symbiotic method of seed germination offered obvious attractions in orchid horticulture.[1] The success of orchid seed germination was achieved by Knudson[15,16] on a medium containing only sugars and mineral salts, and then the non-symbiotic methods are successfully used by many growers.[1] Since then, the non-symbiotic seed germination methods have been very well established in the orchid industry. Now, the non-symbiotic method is applied to orchid seed germination and growth of orchid plants by almost all of the commercial orchid growers. Because the orchid industry became very important in Taiwan in the late 20th century, the author decided to study vesicular arbuscular mycorrhiza since many years ago. Therefore, orchid mycorrhiza is included in my research programs. In Taiwan, it was reported by three orchid mycorrhizal research groups that *Rhizoctonia* spp. could enhance the growth of orchids, including *Bletilla formosana* (Hayata) Schltr. by Su,[17] *Oncidium* spp. by Chu[18] and *Dendrobium* spp. by Lin.[11] This review covers most of the orchid mycorrhizal researches conducted in my laboratory at the National Taiwan University which were aimed at the following: 1) To find out whether the orchid plants grown in green houses are associated with fungi; 2) To test the growth effects of inoculation with orchid mycorrhizal fungi (OMF) in the roots of greenhouse grown orchids; 3) To determine if those inoculated i.e. the mycorrhizal orchid plants were healthier than the non-mycorrhizal control plants; 4) To observe the flowering responses of the mycorrhizal and non-mycorrhizal orchid plants. 5) To understand the growth and developmental response of *Anoectochilus formosanus* Hayata, *Dendrobium* spp., *Haemaria discolor* var. *dawsoniana*, *Phalaenopsis* spp. *Dendrobium* spp., and slipper orchids (*Paphiopedium* spp.) to the inoculation of OMF. Hopefully, the importance of OMF could be recognized and reconsidered by the commercial and amateur orchid growers.

5.2 The Isolation, Purification, and Culturing of Orchid Mycorrhizal Fungi (OMF)

5.2.1 *Isolation, purification, and culturing of OMF*

Wild-grown and greenhouse-cultivated orchid roots are collected and pasteurized using sodium hypochloride solution, then cultured on a 1/6 PDA

Fig. 5.1. Orchid mycorrhizal fungi were isolated from wild grown (**A**), or greenhouse cultivated (**B**), orchid roots (**C**) on 1/6 PDA (**D**).

(potato dextrose agar) medium at 20–30°C (Fig. 5.1). Single hypha is isolated for the pure culture of single fungus. The fungi isolated from the various Taiwan orchid roots are *Acremonium* spp.; *Alternaria* spp.; *Cylindrocarpon* spp.; *Fusarium* spp.; *mycelia sterile; Penecillum* spp.; *Rhizoctonia* spp.; and *Trichoderma* spp. (Table 5.1). *Trichoderma* and *Rhizoctonia* are most frequently isolated from the wild-grown orchid roots. The results showed that greenhouse grown orchids could become mycorrhizal without artificial inoculation.

It was reported that the majority of the fungi associated with orchids are regarded as members of the form-genus *Rhizoctonia*.[1] Thus, the *Rhizoctonia* spp. was isolated (Fig. 5.2). Note the presence of septa in the hyphae, the angle of the branching hyphae (90° or 45°), and the monilioid cells (Fig. 5.2). The isolation results showed that the orchid roots usually harbors more than one fungus. The unifying feature of most isolates of fungi from orchid roots remains their vegetative form, i.e. the features that enables them to be recognized as the *Rhizoctonia* species. At present, the classification of *Rhizoctonia* spp. is based on cytomorphology of hyphae, morphology of cultures, morphology of teleomorphs, affinities for hyphal anastomosis (rejoining of hyphae), the number of nuclei in the cells, and more recently, DNA

Table 5.1. Fungi Isolated from Various Kinds of Healthy Orchid Roots in Taiwan

Fungus Name	Plant Host
Acremonium spp.	*Bletilla, Cleisostoma & Dendrobium* spp.
Alternaria spp.	*Dendrobium* spp.
Cylindrocarpon spp.	*Dendrobium* spp.
Fusarium spp.	*Anoectochilus, Cleisostoma, Dendrobium & Phalaenopsis* spp.
mycelia sterile	*Bulbophyllum, Dendrobium & Paphiopedilum* spp.
Penicillium spp.	*Anoectochilus, Bletilla, Bulbophyllum, Cleisostoma, Dendrobium, Haemaria & Phalaenopsis* spp.
Rhizoctonia spp.	*Anoectochilus, Bletilla, Dendrobium, Haemaria, Spiranthes, Paphiopedilum & Phalaenopsis* spp.
Trichoderma spp.	*Anoectochilus, Paphiopedilum & Phalaenopsis* spp.

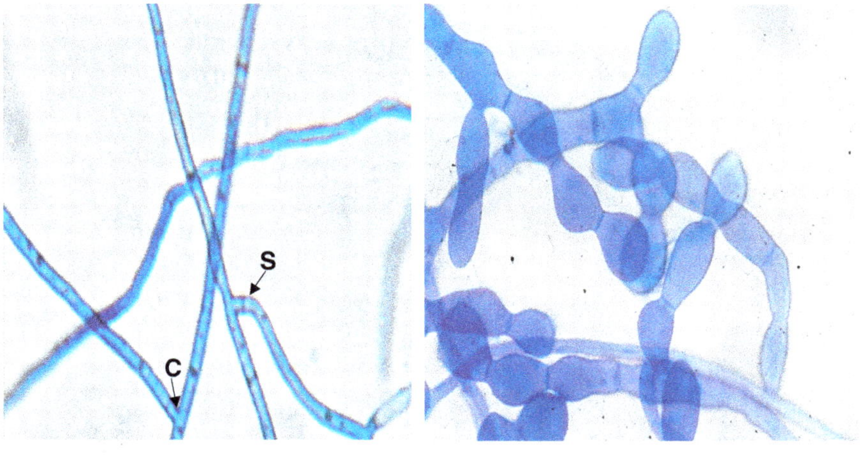

Fig. 5.2. The hyphae with septum (s) and constriction (c) at the base of branching (*left*), and there were monilioid cells appearing in later stages (*right*).

base sequence homology. *Rhizoctonia* is divided into species with binucleate vegetative hyphal cells and those with multinucleate hyphal cells. Several effective staining methods have been developed, such as Aniline blue or Trypan blue, HCl-Geimsa, Orcein, DAPI, and Acridine orange or Safranin O.[19]

5.2.2 *Pathogenecity analysis of the isolated fungi*

The isolates of purified fungi are cultured on PDA; about 1 mm^2 of hyphae are injected under the roots of each seedling, such as cabbage, cucumber, mungbean and rice, which are the major crops in Taiwan (Fig. 5.3). Only those isolates with no harmful symptoms on all the four crop seedlings are chosen as the potential orchid mycorrhizal fungi (POMF). Those isolates of *Rhizoctonia* spp. which show beneficial effects for the growth of orchids are screened as OMF inocula and are named R01–R09. Based on 10 years of inoculation tests, R01, R02, and R04 are shown to be the three isolates with OMF of commercial value. Among them, R01 and R02 are binucleate, while R04 is a multinucleate *Rhizoctonia* spp. Up until now, the identification of R01 and R02 has been uncertain. However, it is known that R04 is *Rhizoctonia solani*, in the AG6,[20] which could enhance seed germination of orchids and was non-pathogenic to the major crops, such as mungbean, cucumber, cabbage, and rice seedlings.

Fig. 5.3. The inoculation of isolated orchid root fungi for pathogenisity test on cucumber, mungbean and rice, respectively.

5.2.3 *The application of orchid mycorrhiza*

5.2.3.1. *Anoectochilus formosanus Hayata and Haemaria discolor var. Dawsoniana*

(A) Seed germination of *Anoectochilus formosanus* Hayata (Fig. 5.4)

For *in vitro* germination, Hyponex # 3 agar medium (containing 3 g of Hyponex # 3, 3 g tryptone, 30 g sucrose and 8 g agar (modified from *Kano, 1968*) was used as a non-mycorrhizal (NM) control and oat meal agar (2.5 g oat meal in 1 L of water, with 8 g of agar, pH 5.7) was used for both mycorrhizal seedlings. For non-symbiotic germination of *H. discolor*, WM agar medium (1 L containing $NaNO_3$, 0.3 g, KH_2PO_4, 0.2 g, $MgSO_4.7H_2O$, 0.1 g, $MgSO_4.7H_2O$, 0.1 g, KCl, 0.1 g, malt extract, 0.1 g, agar, 11.5 g) was used. Results showed that the germination rates were about the same for both treatments.[9,25] However, the growth of seedlings was faster for the symbiotic germination than for the non-symbiotic one (Fig. 5.5).

Jewel's orchids

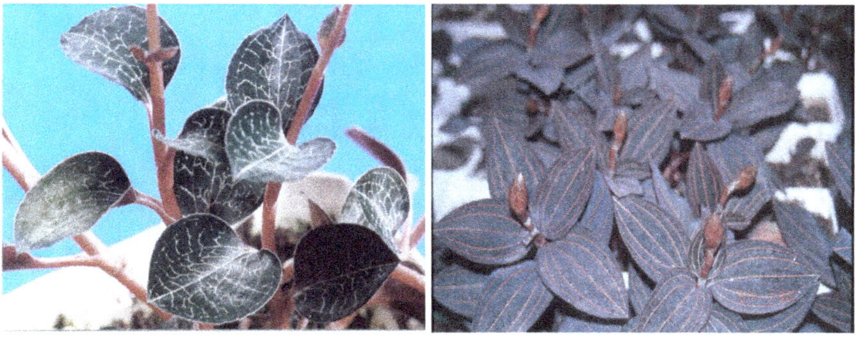

Anoectochilus formosanus* Hayata.** — ***Haemaria discolor* var. *dawsoniana

Fig. 5.4. *Anoectochilus formosanus* was called as "The King of Medicine" by aborigines in Taiwan, used mainly for lung and liver protection, and cancer prevention. *H. discolor* was used for lung protection.

Fig. 5.5. *In vitro* (**A**, **C** & **D**) and *in vivo* germinations (**B**) of *Anoectochilus formosanas* Hayata (**A**, **B**) and *Haemaria discolor* var. *dawsoniana* (**C** & **D**) as affected by orchid mycorrhizal fungi (**A**, +M; **B** & **D**) (Chang, 2006).
NM: non-mycorrhiza; +M: mycorrhiza.

(B) Seedling growth of both *A. formosanus* and *H. discolor in vivo* and *in vitro*

The growth of both seedlings of *A. formosanus*[9,20–24] and *H. discolor*[8] (Fig. 5.6; Refs. 8, 9, 21, 25) was promoted by inoculation of OMF. In most cases, better combinations of plant-OMF are as follows:

A. *formosanus* — R02 or R04;
B. *H. discolor* — R01 or R02.

These results were useful for the production of orchids for both medicinal and ornamental uses.

(C) Component changes of *A. formosanus*

The activities of acid and alkaline phosphatases and SOD (superoxide dismutase), contents of polysaccharides, polyphenols, flavonoids, ascorbic acid, and phosphate were markedly higher in the mycorrhizal tissues of *A. formosanus* than the non-mycorrhizal controls (NMC) (Ref. 9; Table 1). Thus, the mycorrhizal *A. formosanus* should be more effective in their medicinal function.

Currently, wild-grown *A. formosanus* could sell for about four times the price compared with greenhouse-cultivated plants. Actually, most of the wild-grown *A. formosanus* plants were mycorrhizal. The growth of mycorrhizal *H. discolor* var. *dawsoniana* seedlings was almost doubled that of the NMC (Fig. 5.6).

5.2.3.2. *The application of orchid mycorrhizal fungi: Phalaenopsis spp.*

(A) Vegetative growth

Seed germination of many orchids could be greatly enhanced by the presence of OMF.[1,6,12,25–27] Plants of *Phalaenopsis* spp. of various sizes were tested (Fig. 5.7). The growth of most of the *Phalaenopsis* spp. and *Doritaenopsis* spp. cultivars could be enhanced after their transplantation out of glass containers and inoculation with *Rhizoctonia* spp. of OMF.[28]

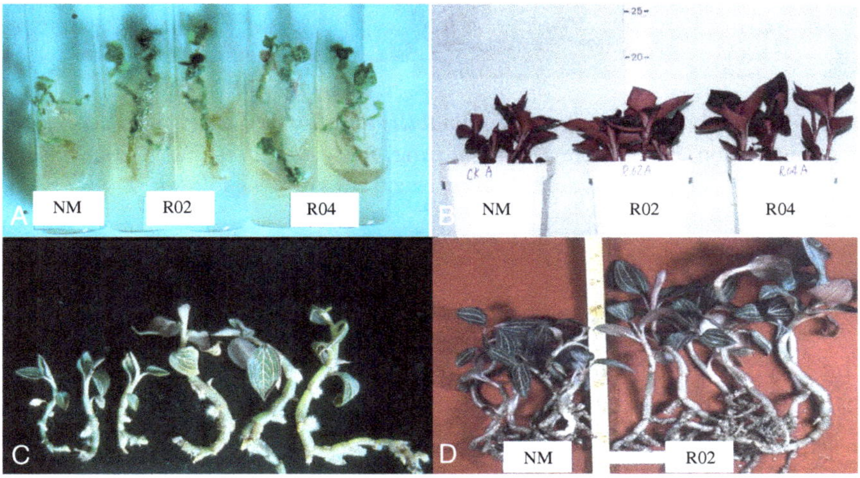

Fig. 5.6. *In vitro* and *ex vitro* seedling growth of *Anoectochilus formosanas* Hayata (**A** & **B**) and *Haemaria discolor* var. *dawsoniana* (**C** & **D**) in oat meal agar (**A** & **C**) and commercial growth media in pots (**B** & **D**) (Chang, 2006).

NM: non-mycorrhiza; R02, R04: mycorrhiza.

Fig. 5.7. In Taiwan, commercially available small, medium and large plants of *Phalaenopsis* orchids are grown in 2.5-, 3.5- and 4.5-inch pots, respectively.

Fig. 5.8. Growth of *Phalaenopsis amabilis* and *Doritaenopsis sp. ex vitro* after the inoculation of orchid mycorrhizal fungi for 4 months (Chang, 2006). Good combinations of host and orchid mycorrhizal fungi are as follows:

Phalaenopsis amabilis: R01, R04.
Doritaenopsis sp.: R02, R04.

The vegetative growth of *Phalaenopsis* spp. could be enhanced by the inoculation of *Rhizoctonia* spp. of OMF (Fig. 5.8); thus, leaf numbers, leaf span, leaf length, and leaf width were enhanced.[21] The chlorophyll contents in some *Phalaenopsis* spp. were greatly increased by the presence of OMF (Fig. 5.9). However, the increase of chlorophyll contents was not consistent for all of the orchids.

(B) Reproductive growth

The reproductive growth (Fig. 5.10) of many *Phalaenopsis* spp. cultivars was enhanced by the presence of OMF, including earlier flowering,

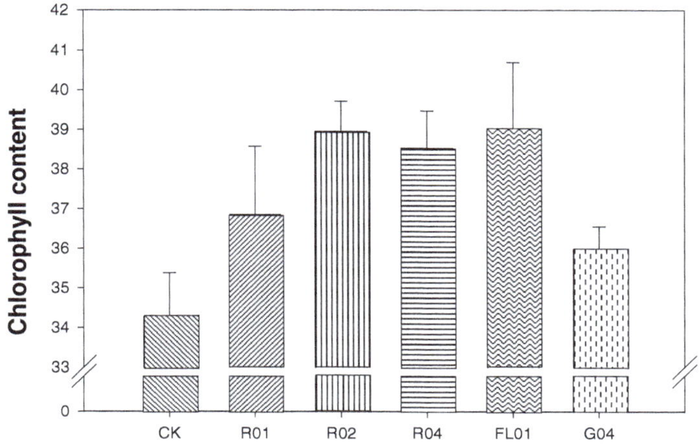

Fig. 5.9. Chlorophyll contents in the leaf of *Dtps.* Luchia Davis × *Dtps.* Taisuco Firebird as influenced by the inoculation of orchid mycorrhizal fungi.

CK: non-mycorrhizal control;
R01, R02, R04, FL01, G04: mycorrhizal plants.

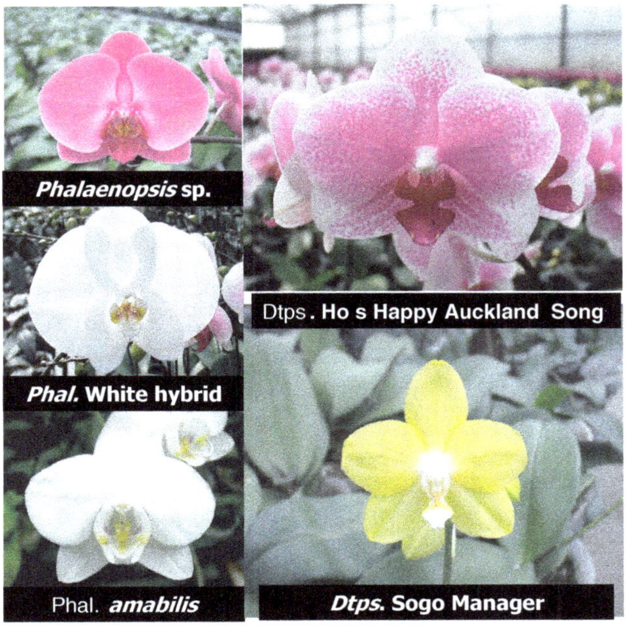

Fig. 5.10. Flowers of tested *Phalaenopsis* spp.

Fig. 5.11. Flower color of *Doritaenopsis* Minho Princess was brighter, as influenced by the inoculation of orchid mycorrhizal fungi (R02), than the non-mycorrhizal control (NM).

increase in flower number, bigger flower diameter, etc.[21] Lan[28] reported better plant growth and a greater percentage of orchid blooming through inoculation with OMF. OMF also resulted in brighter orchid flowers (Fig. 5.11). Thus, it was concluded that both the quantity and quality of orchid flowers were enhanced by OMF.

If, after inoculation with OMF, GA_3 was applied to the *Phalaenopsis amabilis*, a much higher flowering rate under greenhouse conditions could result, without need for the temperature to be lowered.[29] However, if after inoculation with OMF, GA_3 or BA (benzyl adenine), is applied, a greater flower stalk emergence rate could result.[21] Thus, all the results indicated that inoculation with OMF could significantly promote quality and quantity of *Phalaenopsis* spp.

(C) Disease infection

Much less infection was found in the greenhouses of mycorrhizal orchids. In our greenhouse, only about 1% of infection was found for all the mycorrhizal and non-mycorrhizal orchids. In contrast, in the greenhouses of commercial orchid growers, the infection rate was 3% or greater.

One commercial orchid grower in southern Taiwan observed that inoculation of *Phalaenopsis* spp. with *Rhizoctonia* spp. OMF resulted in virus-free orchid plants in a greenhouse, while the non-mycorrhizal orchid plants were all infected with CymMV. Cultivation of virus-free orchid plants is an important goal for most orchid growers. Therefore, research on the ability of OMF to enhance orchid plants to resist infection by CymMV is being conducted and is in progress in our laboratory.

(D) Seasonal infection of OMF

In Taiwan, the percentage of infection of OMF was 100% within one month of inoculation with the three *Rhizoctonia* spp. isolated in Taiwan (R01, R02 and R04), i.e. during spring, summer, and fall. Only in winter season was the percentage of the infection of OMF slightly lower. It was found that the rates of infection of KC 1111 (white flower of *Phalaenopsis* spp.) by the various *Rhizoctonia* spp. varied; some were fast (such as R02, R04, and R19) and some were slow (R01 and R15).

5.2.3.3. *Application of orchid mycorrhizal fungi: Dendrobium spp. and Slipper orchids*

The vegetative growth of *Dendrobium* spp. of orchids — for medicinal uses — could be promoted by inoculation with OMF, such as *Rhizoctonia* spp. and *Cylindricarpon* spp. (Fig. 5.12).[30,31] The growth of slipper orchids could also be promoted by *Rhizoctonia* spp. of OMF.[32,33] After inoculation of OMF (R02 and R04) for six months, leaf length, fresh weight, and survival rates were increased.[33]

5.3 Observations of Orchid Mycorrhizae by Various Microscopies

The orchid mycorrhizae were observed using light microscopy, SEM microscopy, or confocal microscopy. Aniline blue (0.05%) in lactoglycerin was the stain used in light microscopy.[34] SEM samples were prepared by fixing with 2% glutaraldehyde, dehydrating in acetone series, then critical-point drying with liquid carbon dioxide, and finally coated with gold (modified from Ref. 35. Fresh roots of Jewels orchid (*Haemaria* sp.) inoculated with OMF were sectioned using Lancer vibratome series 1000 to obtain sections 30–50 μm in thickness; these were stained with

Fig. 5.12. Growth of *Dendrobium* sp. (**A**), *Oncidium* sp. (**B**), (Chu, 2000) and *Phaphiopedium* sp. (**C**) as affected by orchid mycorrhizal fungi.

(A) NM: non-mycorrhiza; +M: mycorrhiza
(B) CK: non-mycorrhiza; DC, GD, BF2: mycorrhiza
(C) NM: non-mycorrhiza; TR3, TR2, TR1, R04: mycorrhiza

0.05% aniline blue in lactoglycerin for 3 to 4 h (modified from Ref. 34) and were mounted on slides. The stained sections were observed via Leica TCS SP2 laser scanning confocal microscope under 488 and 543 nm, and photos were taken using a computer. The results of light microscopy showed that the hyphae of OMF usually infected root hairs or root epidermal cells in orchids (Fig. 5.13) and to form peletons in the cortex of the roots. In germinating seeds, the fungal hyphae penetrate via the base end of the seed. The hyphae enter the cells and coil into structures called peletons. Germination of the seeds into a portocorm follows. The cells eventually digest the peletons.[2] SEM revealed the three-dimensional structure of the peletons in the cortex cells (Fig. 5.13). Confocal microscopy showed that in the hyphae infection processes,[9] the hyphae passed mainly through intercellular spaces of the root epidermal cells and the few layers underneath, then entered the cortex cells intracellularly to form peletons. According to Burgeff[6] and Hadley,[1] there were two types of OMF infection, i.e. a) Tolypophagy: showing one layer of host cells adjacent to the epidermis and two layers of digestion cells; and b) Ptyophagy: showing two layers of passage cells and the phagocyte layer with hyphae

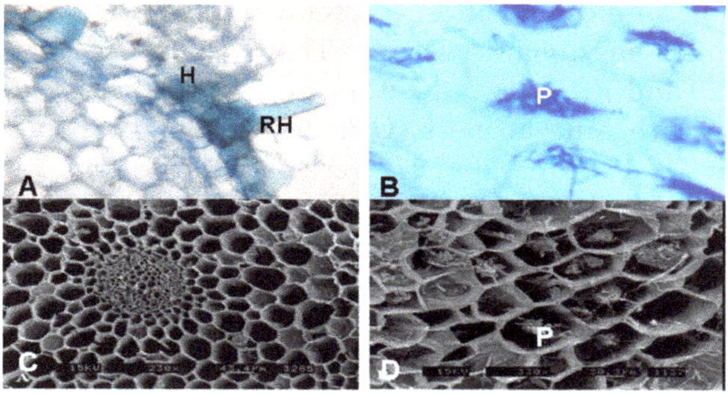

Fig. 5.13. Hyphae (H) of orchid mycorrhizal fungi (OMF) could infect root hair (RH) or orchid root epidermal cells (**A**) and formed peletons (P) in cortex cells (**B**) as stained by aniline blue and revealed by light microscopy. Also, SEM observation of non-mycorrhizal (**C**) and mycorrhizal (**D**) orchid roots showed numerous peletons (P) formed in the mycorrhizal cortex cells, which was a tolypophagy type of infection.

liberating the cytoplasm, which is absorbed into spherical ptysomes. For the infection of three *Rhizoctonia* spp. in several orchid roots (such as *Anoectochilus* sp., *Haemaria* sp., *Dendrobium* sp., *Phalaenopsis* spp., and *Paphiophilium* sp.), only tolypophagy type was observed.

5.4 Identification of OMF by Hyphal Anastomosis Group or RAPD and SSR Markers

Hyphal fusion (anastomosis): Isolates of *Rhizoctonia* are assigned to the anastomosis groups by pairing the isolates with "tester" strains and observing the hyphae for fusion. There are three types of hyphal fusion: perfect fusion, imperfect fusion, and contact fusion.[36] Carling *et al.*[37] reported that of 23 selected mycorrhizal isolates of *Rhizoctonia solani* from an orchid, 20 were members of AG-12 and the rest were members of AG-6.

It is known that our R04 is *Rhizoctonia solani*, which is in the AG6;[20] AG12 also could enhance seed germination of orchids.[38] The pathogenic *Rhizoctonia solani* belong to AG 1, 2, 3, 4, 5, 7, and 8.[38–41]

Rhizoctonia spp. of OMF with multinucleate, were first identified as *Rhizoctonia solani*. Thus, five *R. solani* isolates (R04–7, and R09) from orchids were compared with a total of 19 ATCC test isolates for classifying the anastomosis groups. They were then compared with ATCC 76129 and 76130, two AG-6 isolates, for further grouping.

RAPDs with nine primers (OPD-01, 05–08, 10–12 and 14) and four primers (Y 17, 18, 19 and 27) were used for SSR analysis of OMF. Amplification of ribosomal DNA, sequencing and data analysis were carried out by a commercial company in Taiwan.[20] PCR reactions were based on the analysis techniques recommended by Rafalaski and Tingey.[42] The results of nine OMF calculated from RAPD and SSR were given by Lee.[20] Multinucleate *Rhizoctonia solani* (R04, R05, R06, R07 and R09) all belong to AG-6, which has not shown any pathogenic effects on various kinds of orchids since they were isolated in 1997.

5.5 Specificity of OMF and Hosts

The orchid mycorrhizal fungi (OMF) showed some degree of specificity in its effect on seed germination of orchids (Fig. 5.5). Both binucleate and multi-nucleate *Rhizoctonia* spp. could promote seed germination and growth of orchids. However, usually for seed germination, the binucleate species is more effective. That is why a better combination of plant-OMF should be tested for practical use.

5.6 Important Features for the Successful Application of OMF *In Vitro*

Inoculation of the roots of orchids with *Rhizoctonia* spp. showed that the growth of many kinds of orchids could be thus enhanced (Fig. 5.5, 5.6, 5.8, 5.12 and Table 5.3). *Rhizoctonia* spp. produced a higher germination rate than the multi-nucleated ones.[9] It is very important that, *in vitro*, the growth media for the inoculation be changed to oat meal agar (OMA). If an MS medium or a Hyponex medium for the micropropagation of orchid seedlings or plantlets is used, the growth of OMF would be very fast and is thus harmful to the plantlets or seedlings. Usually these seedlings or plantlets will be covered by the hyphae of OMF and will eventually die (Fig. 5.14). Also the *in vitro* inoculation of OMF was often contaminated by other microorganisms. Therefore, inoculation

Table 5.2. Mycorrhizal *A. formosana* Contained Significantly Higher Enzyme Activities or Component Contents than Non-mycorrhizal Control (modified from Refs. 9 and 21)

Plant Part	Enzyme Activity	Component Contents
Leaf	superoxide dismutase	ascorbic acid, flavonoid, phosphate, polyphenol and polysaccharide
Stem	—	polyphenol and polysaccharide
Root	acid & alkaline phosphatases	ascorbic acid, polyphenol, and polysaccharide

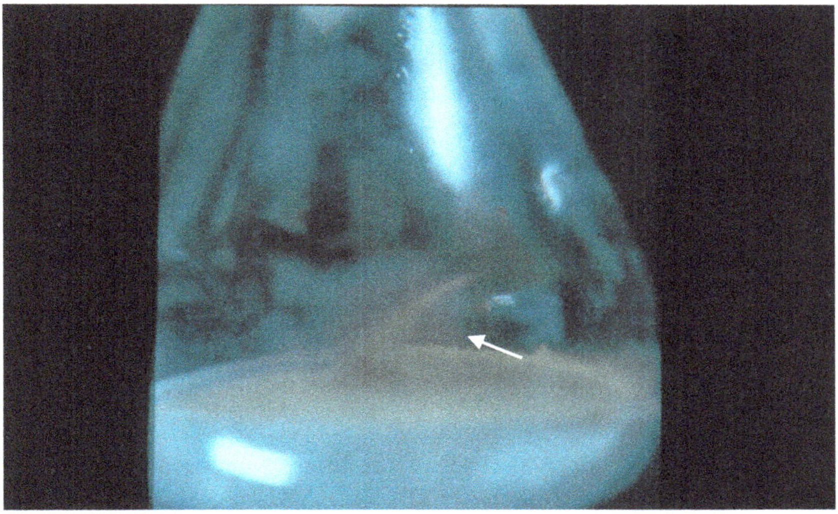

Fig. 5.14. Mass hyphae of orchid mycorrhizal fungus covered and resulted in the death of seedling in MS medium.

with OMF *in vitro* is currently not recommended for mass production of mycorrhizal orchid plants in commercial tissue cultural companies.

There was some degree of plant-OMF specificity for the useful infection of orchid plants by OMF. For example, some *Dendrobium* sp. showed better growth enhancement by infection of R01, while some other *Dendrobium* sp. needed R04 for growth.[30] Therefore, proper combination tests for orchid-OMF should be conducted before mass inoculation of the orchid plants.

Table 5.3. Combinations of Orchid Hosts and *Rhizoctonia* spp. which could result in significant differences in growth of orchids (Chang, 2006)

Host Cultivar	Orchid Mycorrhizal Fungi (OMF)	Vegetative Growth	Reproductive Growth
Anoectochilus formosanas	R02, R04	+	
Haemaria discolor	R01, R02	+	
Dendrobium spp.	R01, R04	+	
Doritaenopsis sp.			
Ho's Happy Auckland 'Song'	R02	+	
Ho's Happy Auckland 'Song'	R02, R01, R04		+
Paphiopedium delenatii	R02, R04	+	
Phalaenopsis spp.			
P. amabilis	R02	+	+
P. Sogo Manager	R01, R02, R04	+	
P. Sogo Manager	R04, R02		+

5.7 Physiological Effects of Orchid Mycorrhizal Fungi and Their Possible Applications

(a) To enhance symbiotic seed germination of orchids *in vitro* and *in vivo*.
(b) To increase the survival rates of the micropropagated plantlets or seedlings *in vivo*. Specificity of OMF-hosts should be considered.
(c) To stimulate nutrient absorption and translocation.
(d) To enhance both vegetative and reproductive growth of orchid plants.
(e) To result in earlier flowering and increase in flower quality.
(f) May help to supply plant hormones (GAs, CK).
(g) To increase chlorophyll contents (photosynthesis might be related).
(h) To increase enzymatic activities (acid and alkaline phosphatases, SOD), and to promote antioxidant abilities by increasing amounts of ascorbic acids, flavonoids, polyphenols, and polysaccharides in *Anoctochilus formosana* Hayata.
(i) To reduce the infection rate and hence less agrichemicals are required.

(j) To combine the use of proper plant growth substances and cool temperature to enhance the production of more flowers as well as to increase flower quality and quantity.
(k) To promote the growth and flowering ability of plants after bare-root shippings.
(l) May be helpful for the restoration of orchids in the wilderness.

Among them, items (b), (d), (e), (i), and (j) were more readily achievable. Better combinations of orchid cultivars and OMF, which could result in significant differences in growth of orchids, are listed in Table 5.2.

It is concluded that various orchid hosts had some degree of specificity for orchid mycorrhizal fungi. Therefore, before large scale applications, tests for proper combinations of orchid-OMF should be conducted. The enhancement of disease resistance of OMF might be the most unique feature that could not be easily replaced by other cultural practices.

References

1. Hadley G. (1982) Orchid Mycorrhiza. In: Arditti J (ed.), *Orchid Biology: Reviews and Perspective* II. Cornell University Press, Ithaca, New York, pp. 83–118.
2. Ian St, G. (2006) Orchids and fungi. New Zealand Native Orchid Group. *Orchid Journal* **66**.
3. Benard N. (1909) L'évolution dans la symbiose. *Ann Sci Nat Bot* **9:**1–196.
4. Burgeff H. (1936) *Samenkeimung der Orchideen.* Fischer, Jena.
5. Burgeff H. (1943) Problematik der Mycorrhiza. *Nautrwiss* **31:**558–567.
6. Burgeff H. (1959) Mycorrhiza of orchids. In: Withner CL (ed.), *The Orchids: A Scientific Survey.* Ronald Press, New York, pp. 361–395.
7. Arditti J. (1967) Factors affecting the germination of orchid seeds. *Bot Rev* **31:**1–97.
8. Chou LC. (1997) Mass propagation of *Haemaria discolor var. dowsoniana* and the use of *Rhizoctonia* spp., Master's thesis. Department of Horticulture, National Taiwan University.
9. Chou LC. (2004) The mycorrhizal physiology and cultivation of *Anoectochilus formosanus* Hayata, *Haemaria discolor* var. *dawsoniana* and their F1 hybrids, Ph.D. dissertation. Institute of Horticulture, National Taiwan University.
10. Deutsch G. (2001) *In vitro*-propagation of *Nigritella* (Orchidaceae-Orchideae) from seeds with the help of mycorrhizal fungi. *Phyton* **41:**111–128.

11. Lin CF. (2002) Effects of orchid mycorrhizal fungi on seed germination and the seedling growth of Dendrobium, Master's thesis. Department of Horticulture, National Chung Hsing University.
12. Sharma J, Zettler LW, Sambeek JWV, *et al.* (2003) Symbiotic seed germination and mycorrhizae of federally threatened *Platanthera praeclara* (Orchidaceae). *Am Midl Nat* **149:**104–120.
13. Takahashi K, Ogiwara I, Hakoda N. (2001) Inoculation of *Habenaria* (Pecteilis) *radiata* with two isolates of orchid mycorrhizal fungi: Germination and growth response to separate, simultaneous, and consecutive inoculation. *Lindleyana* **16:**102–108.
14. Arditti J. (1982) *Orchid Biology Reviews and Perspectives, II.* Cornell Univ. Press.
15. Knudson L. (1922) Nonsymbiotic germination of orchid seeds. *Bot Gaz* **73:**1–25.
16. Knudson L. (1924) Further observations on nonsymbiotic germination of orchid seeds. *Bot Gaz* **77:**212–219.
17. Su MC. (1995) Preliminary study of the symbiosis *Bletilla formosana* in Taiwan, Master's thesis. Department of Forestry, National Taiwan University.
18. Chu JN. (2000) Isolation and inoculation of orchid mycorrhizal fungi and their effects on the seedling growth of *Oncidium Goldiana* × *Onc. Guiena Gold,* Master's thesis. Institute of Tropical Agriculture & International Cooperation.
19. Sneh B, Burpee L, Ogoshli A. (1991) *Identification of Rhizoctonia species.* APS Press, Minnesota, USA.
20. Lee GC. (2001) Identification and improvement of production techniques for *Anoectochilus formosanus* Hayata and orchid mycorrhizal fungi, Ph.D. dissertation. Department of Horticulture, National Taiwan University.
21. Chang DCN. (2006) Research and application of orchid mycorrhizal fungi in Taiwan. *Internatl Hortl Cong (IHC) 2006,* Aug. 13–19, Seoul, Korea.
22. Li HMC. (2001) Physiology of orchid mycorrhizal fungi (*Rhizoctonia* spp.), inoculum production and their effects on the growth of *Anoectochilus formosanus* Hayata, Master's thesis. Department of Horticulture, National Taiwan University.
23. Sheu CL. (1994) The effect of mycorrhizae on the growth of Carrot plantlet derived from somatic embryo and *Anoectochilus formosanus* Hayata tissue culture plantlet, Master's thesis. Department of Horticulture, National Taiwan University.
24. Tsai CY. (1997) Effects of *Rhizoctonia* spp, and temperature on the growth of *Anoectochilus formosanus* Hayata plantlets, Master's thesis. Department of Horticulture, National Taiwan University.
25. Chang DCN, Chou LC. (2001) Seed germination of *Haemaria discolor* var. *dawsoniana* and the use of mycorrhizae. *Symbiosis* **30:**29–40.

26. Uetake Y, Farquhar ML, Peterson RL. (1997) Changes in microtubules arrays in symbiotic orchid protocorms during fungal colonization and senescence. *New Phytol* **135:**701–709.
27. Zettler LW, Hofer CJ. (1998) Propagation of the little club-spur orchid (*Platanthera clavellata*) by symbiotic seed germination and its ecological implications. *Environ Exp Bot* **39:**189–195.
28. Lan IC. (2001) Effects of orchid mycorrhizal fungi on the growth and development of *Phalaenopsis* and *Paphiopedium* spp., Master's thesis. Department of Horticulture, National Taiwan University.
29. Wang YF. (2005) Effects of orchid mycorrhizal fungi and plant growth substances on the growth and flowering of *Phalaenopsis* spp., Master's thesis. Department of Horticulture, National Taiwan University.
30. Kang CW. (2004) Effects of orchid mycorrhizal fungi and plant growth substances on the growth and development of medical *Dendrobium* spp., Master's thesis. Department of Horticulture, National Taiwan University.
31. Tseng CJ. (2001) Isolation, identification and culture of *Dendribium* mycorrhizal fungi and their effects on the growth of *Dendrobium* spp., Master's thesis. Department of Horticulture, National Taiwan University.
32. Tsai LC. (2003) Effects of orchid mycorrhizal fungi and plant growth substances on the growth and development of *Paphiopedilum* spp., Master's thesis. Department of Horticulture, National Taiwan University.
33. Chen CC. (2005) Effects of orchid mycorrhizal fungi with plant growth substances on the growth and development of *Paphiopedilum* spp., Master's thesis. Department of Horticulture, National Taiwan University.
34. Koske RE, Gemma JN. (1989) A modified procedure staining roots to detect VA mycorrhizas. *Mycol Res* **92:**486–488.
35. Porter KR, Kelly D, Andrews PM. (1972) The preparation of cultured cells and soft tissues for scanning electron microscopy. *Proc Fifth Ann Stereoscan Colloq* (1972). Kent Cambridge Corp., 19 pp.
36. Matsumoto T, Yamamoto W, Hirane S. (1932) Physiology and parasitology of the fungi generally referred to as Hypochnus sasakii Shirai. I. Differentiation of the strains by means of hyphal fusion and culture in differential media. *J Soc Trop Agric* **4:**370–388.
37. Carling DE, Pope EJ, Brainard KA, Carter DA. (1999) Characterization of mycorrhizal isolates of *Rhizoctonia solani* from an orchid, including AG-12, a new anastomosis group. *Phytopathology* **89:**942–946.
38. Pope J, Carter DA. (2006) Phylogenetic placement and host specificity of mycorrhizal isolates belonging to AG-6 and AG-12 in the *Rhizoctonia solani* species complex. *Mycologia* **93:**712–719.
39. Harikrishnan R, Yang XB. (2002) Effects of herbicides on root rot and damping-off caused by *Rhizoctonia solani* in glyphosate-tolerant soybean. *Plant Dis* **86:**1369–1373.

40. Priyatmojo A, Escopalao VE, Tangonan NG, *et al*. (2001) Characterization of a new subgroup of *Rhizoctonia solani anastomosis* group 1 (AG-1-ID), causal agent of a necrotic leaf spot on coffee. *Phytopathology* **91:**1054–1061.
41. Stepniewska-Jarosz S, Manka M, Asiegbu FO. (2006) Studies on anastomosis group of isolates causing disease in two forest nurseries in Poland. *Pathology* **36:**97–109.
42. Rafalski JA, Tingey SV. (1993) Genetic diagnostics in plant bredding: RAPDs, microsatellites and machines. *TIG* **9:**275–280.

Chapter 6

Analysis of the Orchid Genome Size Using Flow Cytometry

Tsai-Yun Lin[*,†] *and Hsiao-Ching Lee*[†]

Flow cytometry is an efficient and reliable method for the estimation of nuclear DNA contents. Nuclei suspension from orchid plant is often contaminated with a high level of crystalline calcium oxalate that blocks the fluidics system of the flow cytometer. A simple and highly efficient protocol enables isolation of intact nuclei from recalcitrant plant tissues containing high levels of polysaccharides, calcium oxalate crystals, and other metabolites. This method was applied to determine the genome sizes of *Phalaenopsis* sp. using *Pisum sativum* Minerva Maple as the standard. The occurrence of endoreduplication in orchids during organ development and the influences of environmental perturbations are also discussed.

6.1 Introduction

Interspecific hybridization and chromosome doubling are the techniques often applied to produce new cultivars of orchids.[1] A better understanding of the karyotypes and DNA contents of orchid will aid in the development of new cultivars of high quality. Intraspecific variations in genome size occur due to environmental influences, common chromosome polymorphism, and spontaneous aberration. However, the DNA content per genome is fairly constant, both between cells of an

*Corresponding author.
†*Institute of Bioinformatics and Structural Biology and Department of Life Science, National Tsing Hua University, Hsinchu, Taiwan.*

individual and between different individuals of the same species.[2,3] An accurate determination of genome size provides basic information for breeders and molecular geneticists. Interspecific comparison of nuclear DNA amounts is also useful in cytotaxonomic and evolutionary studies. The classical methods for the estimation of genome sizes are based on the determination of the phosphate content in the DNA backbone of total DNA isolated from a defined number of cells or on re-association kinetics of high molecular weight genomic DNA. More recent techniques employ DNA-specific fluorescent dyes in flow cytometry analysis.

Cell polyploidy is often due to endoreduplication, when one or several rounds of DNA synthesis occur in the absence of mitosis or cytokinesis, resulting in an increase in the genomic DNA content.[4] Endoreduplication is an important component of organ development and is often related to plant cell size. The number of endoreduplication cycles is tissue-specific and is characteristic of the stage of development.[5]

6.2 Flow Cytometric Methods

6.2.1 *Flow cytometers*

Flow cytometry has proven to be an efficient and reliable method for analyzing plant genomes. The principle of flow cytometry is that single particles suspended within a stream of liquid are interrogated individually in a very short time as they pass through a light source focused at a very small region. The optical signals generated are mostly spectral bands of light in the visible spectrum, which represent the detection of various chemical or biological components, mostly fluorescence.[6]

6.2.2 *Applications of flow cytometry*

DNA flow cytometry has become a popular method for ploidy screening, detection of mixoploidy and aneuploidy, cell cycle analysis, assessment of the degree of polysomaty, determination of reproductive pathways, and estimation of absolute DNA amount or genome size.[7] Plant scientists have used flow cytometry to determine the nuclear DNA content of a wide variety of species, such as *Arabidopsis*[8]; cucumber[9]; tomato[10]; orchids[11,12]; several succulents[13]; and endosperm in maize.[14] In addition, flow cytometry has been used to analyze the cell cycle, sort protoplasts and chromosomes, and examine various cellular parameters.[15]

6.3 Estimation of Orchid Genome Size

6.3.1 *Preparation of orchid nuclei suspensions*

There is no general procedure that works with all plants, although several protocols have been described for nuclei isolation.[16–18] Most protocols use young tissues to avoid the high concentration of starch, polysaccharides, calcium oxalate crystals, and other metabolites found in old tissues. Orchid plant nuclei suspension often contains a high level of crystalline calcium oxalate that blocks the fluidics system of the flow cytometer. To avoid this problem, a cotton column was designed and polyvinyl pyrrolidone-40 (PVP-40) was added to the buffer to remove phenolic impurities and cytoplasmic compounds from plant nuclei, making the suspension suitable for flow cytometry. This simple and highly efficient protocol enables the isolation of intact nuclei from plant tissues containing high levels of polysaccharides, calcium oxalate crystals, and other metabolites.[19]

6.3.1.1 *Solution and reagents*

- $MgSO_4$ buffer: 10 mM $MgSO_4 \cdot 7H_2O$, 50 mM KCl, 5 mM HEPES [*N*-2-hydroxyethylpip erazine-*N′*-2-ethanesulfonic acid], pH 8.0.
- Extraction buffer A: $MgSO_4$ buffer with 1% polyvinyl pyrrolidone-40, 6.5 mM dithiothreitol, 0.25% Triton X-100; stored at 4°C.
- Extraction buffer B: $MgSO_4$ buffer with 6.5 mM dithiothreitol, 0.25% Triton X-100, 0.2 mg ml^{-1} propidium iodide (Calbiochem, La Jolla, Calif.), 1.25 $\mu g\ ml^{-1}$ RNase (DNase-free); prepared on ice just prior to use.

6.3.1.2 *Column for nuclei filtration*

To make the column for nuclei filtration, cosmetic cotton made of 100% cotton is cut into 5 × 1-cm pieces moistened with $MgSO_4$ buffer and loosely rolled into a cylindrical shape. The cotton ball is gently inserted into the middle of a 5 ml pipette tip (Fig. 6.1). The cotton ball filters out

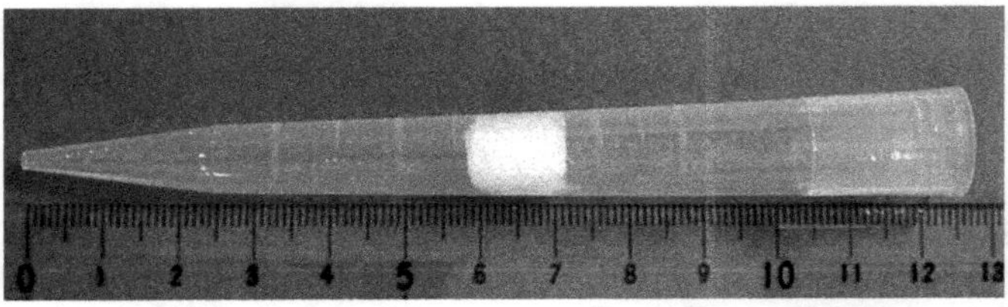

Fig. 6.1. The cotton column for filtration of plant nuclei. Cosmetic cotton is cut into 5 × 1-cm pieces, wetted with $MgSO_4$ buffer and rolled into a cylindrical shape loosely, then inserted into a 5 ml pipette tip to form a filtration column.[19] (Photo from Lee *et al.*, courtesy of the ISPMB abd PMBR.)

the particles on the basis of differences in physical structure of the particles, and the position of the cotton ball affects the efficiency of filtration and yield of nuclei. Users may shift the position of the cotton ball up or down to find the optimal pore size for different plant materials. Prior to use, the column must be rinsed with 10 ml or more of $MgSO_4$ buffer solution to remove floating cotton fibers.

6.3.1.3 *Nuclei isolation*

While on ice, 50 to 300 mg of fresh tissue is sliced into thin strips of less than 0.5 mm with a sharp razor blade in a glass petri dish containing 1 ml of ice-cold extraction buffer A. Two ml of ice-cold extraction buffer A is added to the chopped tissue and gently shaken for 1 min on a shaker or by hand. The nuclei extract is filtered into a 15-ml Falcon tube through the cotton column. The homogenate is transferred to the cotton column and nuclei are eluted with a 7 ml ice-cold extraction buffer A. The homogenate is centrifuged at 120 *g* for 10 min at 4°C. The supernatant is then carefully decanted and the pellet gently suspended in 400 μl of extraction buffer B, which is then incubated in the dark at 37°C for 15 min before flow cytometry analysis.

6.3.1.4 *Flow cytometry analysis*

Flow cytometry analysis lacks the intrinsic ability to deal with cell aggregates or other large particles, thus requiring the isolation of intact nuclei. Isolating intact nuclei from plant tissues is not easy, especially when young tissues are unavailable. Fully developed, mature organs are usually heavily loaded with polysaccharides, calcium oxalate crystals, and

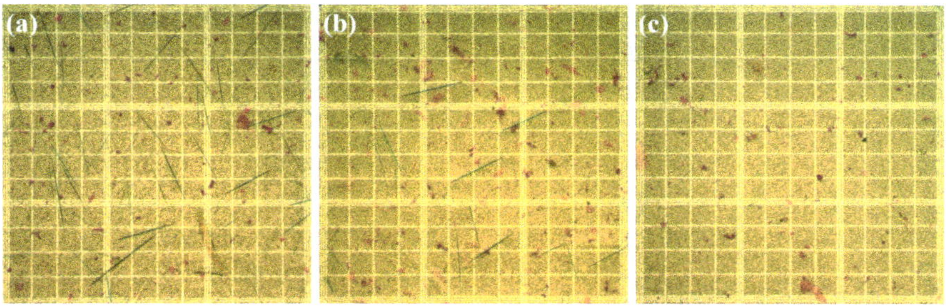

Fig. 6.2. Effect of the cotton column on plant nuclei isolation confirmed by microscopy. Samples prepared from mature leaves of *Phalaenopsis aphrodite* subsp. *formosana* and stained with carmine showing nuclei were examined under a microscope (Axioskop, Zeiss, Germany) with a hemacytometer. (a) The suspension without filtration contains a large amount of calcium oxalate crystals (needle-shaped) that clog the tube of flow cytometer. (b) The suspension filtered through a 30-μm nylon mesh still contains many calcium oxalate crystals and other residues of cells. (c) The suspension filtered through the cotton column significantly reduced contaminants.[19] (Photo from Lee *et al.*, courtesy of the ISPMB and PMBR.)

other metabolites that decrease the purity of intact nuclei. These contaminants, such as calcium oxalate crystals, have smaller diameters than the nuclei and are difficult to remove by the small pore size (30–50 μm) nylon mesh generally used in other protocols.[16–18] The cotton column forms meshed network to retard the irregular-shaped contaminants, and the addition of PVP-40 to the buffer removes phenolic impurities. Microscopy confirmed the purity of the isolated nuclei (Fig. 6.2). The disadvantage of this protocol is a 50% decrease in nuclei yield. To overcome this problem, the amount of tissue used needs to be increased for nuclei extraction. This cotton column is particularly effective on recalcitrant plant tissues and allows precise measurement using flow cytometry.

6.3.2 *Fluorescent staining of nuclear DNA*

Propidium iodide (PI) and 4′-6-diamidino-2-phenylindole (DAPI) are commonly used fluorochromes. The choice of fluorochromes is primarily determined by the excitation source available. PI is excited by visible light with an absorbancy maximum at 490 nm, while DAPI is excited by UV light at 350 nm.[20] However, the two dyes have quite different stain reactions. PI intercalates between base pairs of double-stranded DNA and RNA with little or no base specification,[21] while DAPI is a nonintercalating stain that binds preferentially and in a complex manner to A-T base regions.[22] Highly significant differences were detected in plant nuclear DNA contents obtained using DAPI and PI and this cast doubt on the reliability of base preference fluorochromes.[23] PI-based flow cytometry produced results consistent with those based on Feulgen microspectrophotometry; DAPI was suggested not to be used as the sole basis for DNA content comparisons.[20] PI was recommended as the fluorochrome of choice for flow cytometric determination of plant DNA content. DAPI should be used only if the estimated DNA value is corroborated by using a second stain that has no bias for AT- or GC-rich sequences within genomes.[20]

6.3.3 *DNA reference standards*

Using proper reference standards for flow cytometric determination of DNA contents is important. Most of the published DNA values for plants have been calibrated against chicken erythrocytes. However, this may result in significant error in estimated DNA contents, as there is considerable reported variations in DNA content among chicken lines.[24] Price *et al.*[25] suggested that for technical reasons a plant standard is better for estimating the DNA content of plants. Five species are recommended by Johnston *et al.*[20] as an initial set of international standards for plant DNA content determinations: *Sorghum bicolor* cv. Pioneer 8695 (2C = 1.74 pg); *P. sativum* cv. Minerva Maple (2C = 9.56 pg); *Hordeum vulgare* cv. Sultan (2C = 11.12 pg); *Vicia faba* (2C = 26.66 pg); and *Allium cepa* cv. Ailsa Craig (2C = 33.55 pg). It is recommended that one of the reference standards be chosen for determination of DNA content of plant genome.

6.4 Nuclei DNA Contents of Orchids

6.4.1 *Nuclei DNA contents of Phalaenopsis sp.*

Chromosome numbers for the orchids are usually examined and stained by basic fuchsin and acetic orcein. There is a 6.07-fold variation in genome size within 18 *Phalaenopsis* species, ranging from 2.74 pg/2C for *P. sanderiana* to 16.61 pg/2C for *P. parishii* (Table 6.1).[11]

Table 6.1. Nuclear DNA Contents of *Phalaenopsis* sp. Determined by Flow Cytometry of Propidium Iodide Attained Nuclei using *P. sativum* Minerva Maple as the Standard[11]

Genus	Section[a]	Species[a]	TSRI Clone[b]	Mean (pg/2C)
Phalaenopsis	*Phalaenopsis*	*P. aphrodite* Rchb.f.[c]	W01–38	2.80
		P. sanderiana Rchb.f.	W36–01	2.74
		P. stuartiana Rchb.f.	W40–04	3.13
	Stauroglottis	*P. equestris* (Schauer) Rchb.f.	W09–51	3.37
	Amboinenses	*P. amboinensis* J. J. Smith	W02–07	14.36
		P. amboinensis J. J. Smith	W02–08	14.50
		P. gigantea J. J. Smith	S82–314	5.28
		P. micholitzii Rolfe	W27–05	6.49
		P. venosa Shim & Fowl.	W42–01	9.52
	Zebrinae	*P. fasciata* Rchb.f.	W10–01	6.56
		P. lueddemanniana Rchb.f.	W23–02	6.49
		P. modesta J. J. Smith	W28–06	5.15
		P. mariae Burb. ex Warn. & Wms.	W26	6.48
		P. pulchra (Rchb. f.) Sweet	W33–04	6.37
		P. sumatrana Korth. & Rchb.f.	W41–03	6.62
		P. bellina (Rchb. f.) Cristenson[d]	S82–429	15.03
	Polychilos	*P. cornu-cervi* (Breda) Bl & Rchb.f.	W08–01	6.44
		P. mannii Rchb.f.	W25–06	13.50
		P. mannii Rchb.f.	W25–07	13.61
	Parishianae	*P. parishii* Rchb.f.	W32	16.61

[a]Classification according to Sweet (1980).
[b]Source of plant material: Taiwan Sugar Research Institute (TSRI).
[c]Chromosome number are $2n = 2x = 38$.
[d]Nomenclature according to Christenson and Whitten (1995).

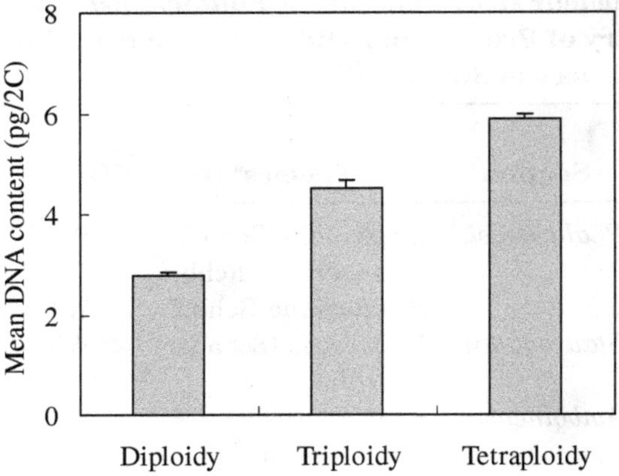

Fig. 6.3. Nuclear DNA contents of diploidy, triploidy, and tetraploidy of *P. aphrodite* determined by flow cytometry of propidium iodide stained nuclei. The DNA contents (mean ± SE) of diploidy, triploidy, and tetraploidy of *P. aphrodite* are 2.8 ± 0.06, 4.52 ± 0.16 and 5.89 ± 0.10 pg/2C, respectively.[11] Nuclei extracted from *P. sativum* Minerva Maple were used as the standard.

Chromosome numbers of different TSRI clones of *P. aphrodite* were confirmed to be $2n = 2x = 38$ for W01-38; $2n = 3x = 57$ for W01-41; and $2n = 4x = 76$ for W01-22. The nuclear DNA contents of diploidy, triploidy and tetraploidy of *P. aphrodite* had a ratio near 2:3:4, corresponding to their somatic chromosome numbers, 38, 57 and 76 (Fig. 6.3). The linear relationship between nuclear DNA contents and chromosome numbers of these three clones indicated the accuracy of flow cytometry.

6.4.2 *Intraspecific variations in genome size*

No variation in DNA content was found between the two different *P. amboinesis* clone, W02–07 (14.36 pg/2C) and W02–08 (14.50 pg/2C) or between the two different *P. mannii* clones, W25–07 (13.50 pg/2C) and W25–06 (13.61 pg/2C) (Table 6.1). Our DNA 2C-values of 3.37 pg/2C for *P. equestris* and 6.49 pg/2C for *P. lueddemanniana* are smaller than that previously reported (5.53 pg/2C for *P. equestris* and 8.65 pg/2C for *P. lueddemanniana*).[26] These differences could have resulted from intraspecific variations.[11]

6.5 Endoreduplication Occurrence in Orchid

6.5.1 *Phenomenon of endoreduplication*

Cell polyploidy is often due to endoreduplication, when one or several rounds of DNA synthesis occur in the absence of mitosis or cytokinesis, resulting in an increase in the genomic DNA content.[4,27–29] Endoreduplication is common in animal and plant cells, especially in those undergoing differentiation and expansion.[30–32] Endoreduplication in the endosperm and cotyledons of developing seeds is well documented, but it also occurs in many tissues throughout the plant. An increase in ploidy may be part of the differentiation of a single cell, as it is in *Arabidopsis* trichomes,[33–35] and may also be related to cell size.[36] Cell size in yeast increases with increasing ploidy.[37] In plants, increases in ploidy level are related to increases in nuclear volume and cell size. Cells that have undergone endoreduplication are larger than comparable cells that have not.[5,38]

Endoreduplication is widespread in orchid plants. Sturdier flowers and better forms may accompany an increase in ploidy in orchids. Although chromosome doubling is a technique for breeders to obtain new cultivars, little is known regarding the mechanism of endoreduplication in orchids. Use of flow cytometry in determining polyploidy is fast and makes determining polyploidy easier.

6.5.2 *Occurrence of endoreduplication during organ development*

The frequency of nuclei at different ploidy levels differs in various organs and is correlated with development. Older tissues often possess nuclei of higher ploidy levels than younger ones within the same plant. For example, in *Cucumis sativus*, ploidy levels were described as a continuous process during the developmental stages from seed to flowering plants.[9] The distribution of nuclei at different ploidy levels also changes with the developmental stages as determined by flow cytometry. The number of endoreduplication cycles appears to be controlled by developmental signals.[27] Nevertheless, direct evidence for the link between development and the ploidy level in various organs and tissues of plant still remains to be determined.[4]

In *P. aphrodite* and *P. equestris*, endoreduplication stopped once the flower was fully expanded,[12,39] and ploidy levels stabilized, similar to leaves in *Arabidopsis*,[8] cucumber,[9] and several succulents.[13] Floral buds and opened flowers in an inflorescence had different frequencies of endoreduplication. The distribution of endopolyploidy also varied among the different parts of a flower (Fig. 6.4). The proportion of endoploidy in cells increased with the increasing size of a floral bud. A similar pattern of polysomaty is found in the fully opened flowers. Our results showed that the younger floral buds have higher proportions of 2C cells. To describe the relationship between endoreduplication and cell growth, a model was designed combining a logistic growth model with an endoreduplication model of Schweizer *et al.*[14] Our results indicate that cells with higher C values had lower transition rates and less potential for further endoreduplication, and when cessation of endoreduplication occurred, the flower fresh weight also stopped increasing. The average cell fresh weight was positively correlated to the average C value, suggesting that endoreduplication in *Phalaenopsis* is a contributing factor to cell growth.[39]

The *Arabidopsis* plant-specific cyclin-dependent kinase CDKB1;1 and transcription factor E2Fa-DPa control the balance between mitotic division and endoreduplication.[40] Plants which overexpressed a dominant negative allele of CDKB1;1 were found to undergo enhanced endoreduplication, demonstrating that the CDKB1;1 activity is required to inhibit the endocycle. In E2Fa-DPa–over expressing plants, DNA replication is strongly activated. Plant organ cell number and cell size might be controlled by modulation of genes involved in cyclin D-CDK inhibitor interaction. Plant CDK inhibitors, ICK1/KRP1 can move between cells and can inhibit entry into mitosis.[41]

6.5.3 *Endoreduplication influenced by environmental perturbations*

Environmental perturbations can influence endoreduplication. For example, both DNA endoreduplication and mitosis in maize endosperm cells were significantly correlated with decreased fresh weight and nuclei number after exposure to 35°C for 4 to 6 days.[42] In *Arabidopsis*, reductions in light intensity and soil water content reduces the extent of endoreduplication.[42] The hypocotyls cells go through several cycles of endoreduplication, and the process is differentially regulated in the light and in the dark.[44] Activated phytochrome specifically inhibits a third endoreduplication cycle in developing hypocotyls cells. The 16C

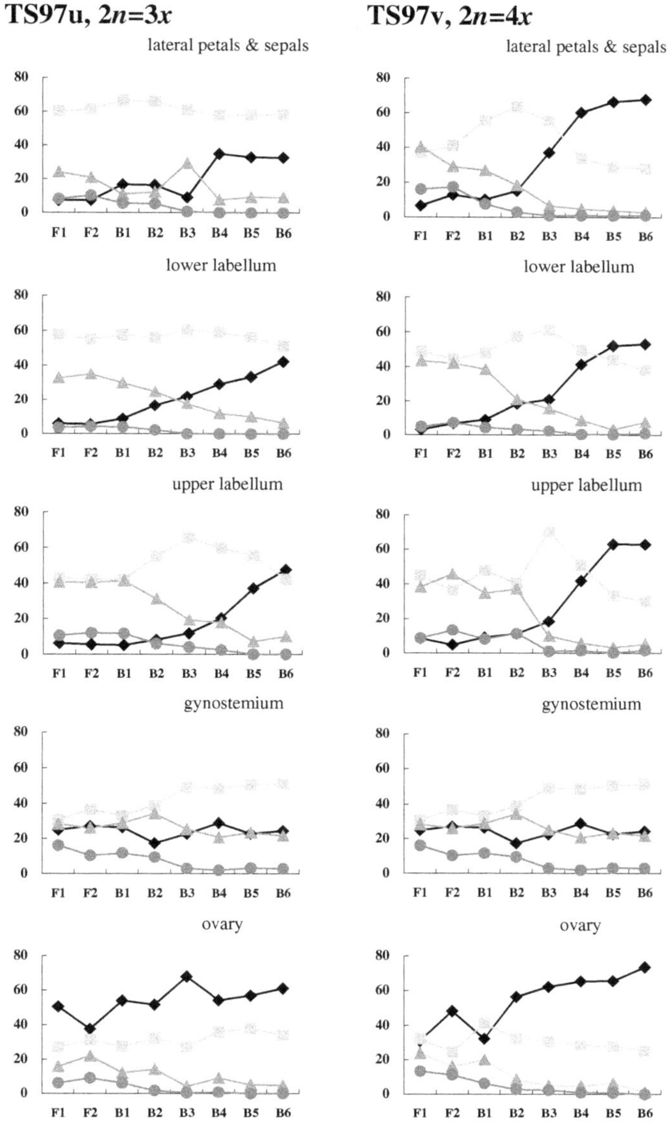

Fig. 6.4. Frequencies of cells with various C-values in different parts of flower. The flowers of TS97u ($2n = 3x$) and TS97v ($2n = 4x$) in a florescence were numbered from the bottom to the top, with the oldest flower as number one. The ploidy distribution of two inflorescences was recorded. The fully opened flowers had a similar pattern of polysomaty (◆: 2C, ■: 4C, ▲: 8C, ●: 16C).

nuclei were present in the hypocotyls of *phytochrome* (*phy*) *A* mutants, but not in that of wild type *Arabidopsis* plants under continuous far red light.[45] The salt treatment (150 mM NaCl) induced chromosome endoreduplication during the differentiation of cells in the root cortex of *Sorghum bicolor* cv. 610 plants, but not in the cells of the root vascular cylinder or the leaves. In another *S. bicolor* genotype DK 34-Alabama, noncompetent for salt adaptation, the same salt treatment did not induce chromosome endoreduplication in the root cortex cells. Endopolyploidy may be considered as a part of the adaptive response of *S. bicolor* competent genotypes to salinity.[46] The number of endoreduplication cycles appears to be under the control of developmental and environmental signals.

Acknowledgments

We thank AH Marhkart for comments on the manuscript and the National Science Council, Republic of China for grant NSC 92-2317-B-007-001 and Council of Agriculture, Republic of China for grants 94AS-5.2.1-ST-a1 and 95AS-6.2.1-ST-a1.

References

1. Arditti J. (1997) *Fundamentals of Orchid Biology*. Wiley, New York.
2. Greilhuber J. (1998) Intraspecific variation in genome size: A critical reassessment. *Ann Bot* **82(Supp. A):**27–35.
3. Greilhuber J, Obermayer R. (1997) Genome size and maturity group in *Glycine max* (soybean). *Heredity* **78:**547–551.
4. Joubès J, Chevalier C. (2000) Endoreduplication in higher plants. *Plant Mol Biol* **43:**735–745.
5. Kudo N, Kimura Y. (2002) Nuclear DNA endoreduplication during petal development in cabbage: Relationship between ploidy levels and cell size. *J Exp Bot* **53:**1017–1023.
6. Gary W, Gary LB. (2004) *Encyclopedia of Biomaterials and Biomedical Engineering*. Marcel Dekker, New York.
7. Dolezel J, Bartos J. (2005) Plant DNA flow cytometry and estimation of nuclear genome size. *Ann Bot* **95:**99–110.
8. Galbraith DW, Harkins KR, Knapp S. (1991) Systemic endopolyploidy in *Arabidopsis thaliana*. *Plant Physiol* **96:**985–989.

9. Gilissen LJW, van Staveren MJ, Creemers-Molenaar J, Verhoeven HA. (1993) Development of polysomaty in seedlings and plants of *Cucumis sativus* L. *Plant Sci* **91:**171–179.
10. Smulders MJM, Rus-Kortekaas W, Gilissen LJW. (1994) Development of polysomaty during differentiation in diploid and tetraploid tomato (*Lycopersicon esculentum*) plants. *Plant Sci* **97:**53–60.
11. Lin S, Lee HC, Chen WH, *et al.* (2001) Nuclear DNA contents of *Phalaenopsis* sp. and *Doritis pulcherrima*. *J Amer Soc Hort Sci* **126:**195–199.
12. Yang M, Loh CS. (2004) Systemic endopolyploidy in *Spathoglottis plicata* (Orchidaceae) development. *BMC Cell Biol* **5:**33.
13. de Rocher EJ, Harkins KR, Galbraith DW, Bohnert HJ. (1990) Developmentally regulated systemic endopolyploidy in succulents with small genomes. *Science* **250:**99–101.
14. Schweizer L, Yerk-Davis GL, Phillips RL, *et al.* (1995) Dynamics of maize endosperm development and DNA endoreduplication. *Proc Natl Acad Sci USA* **92:**7070–7074.
15. Brown SC, Bergounioux C. (1989) Plant flow cytometry. In: Yen A (ed.), *Flow Cytometry: Advanced Research and Clinical Applications, Vol. 2.* CRC Press, Boca Raton, Florida, pp. 195–220.
16. Galbraith DW, Harkins KR, Maddox JM, *et al.* (1983) Rapid flow cytophotometric analysis of the cell cycle in intact plant tissues. *Science* **220:**1049–1051.
17. Arumuganathan K, Earle ED. (1991) Estimation of nuclear DNA content of plants by flow cytometry. *Plant Mol Bio Rep* **9:**229–233.
18. Ulrich I, Ulrich W. (1991) High-resolution flow cytometry of nuclear DNA in higher plants. *Protoplasma* **165:**212–215.
19. Lee HC, Lin TY. (2005) Isolation of plant nuclei suitable for flow cytometry from recalcitrant tissue using a filtration column. *Plant Mol Biol Rep* **23:**53–58.
20. Johnston JS, Bennett MD, Rayburn AL, *et al.* (1999) Reference standards for determination of DNA content of plant nuclei. *Amer J Bot* **86:**609–613.
21. Properi E, Giangare MC, Bottiroli G. (1991) Nuclease induced DNA structural changes assessed by flow cytometry with the intercalating dye propidium iodide. *Cytometry* **12:**323–329.
22. Godelle B, Cartier D, Marie D, *et al.* (1993) Heterochromatin study demonstrating the non-linearity of fluorometry useful for calculating genomic base composition. *Cytometry* **14:**618–626.
23. Dolezel J, Sgorbati S, Lucretti S. (1992) Comparison of three DNA fluorochromes for flow cytometric estimation of nuclear DNA content in plants. *Physiol Plantarum* **85:**625–631.
24. Bennett MD, Leitch I. (1995) Nuclear DNA amounts in angiosperms. *Ann Bot* **76:**113–176.

25. Price HJ, Bachmann K, Chambers KL, Riggs J. (1980) Detection of intraspecific variation in nuclear DNA content in *Microseris douglasii*. *Bot Gaz* **141:**195–198.
26. Jones WE, Kuehnle AR, Arumuganathan K. (1998) Nuclear DNA content of 26 orchid (Orchidaceae) genera with emphasis on *Dendrobium*. *Ann Bot* **82:**189–194.
27. Traas J, Hülskamp M, Gendreau E, Höfte H. (1998) Endoreduplication and development: Rule without dividing? *Curr Opin Plant Biol* **1:**498–503.
28. Grafi G, Larkins BA. (1995) Endoreduplication in maize endosperm: Involvement of M phase-promoting factor inhibition and induction of S phase-related kinases. *Science* **269:**1262–1264.
29. Grafi G. (1998) Cell cycle regulation of DNA replication: The endoreduplication perspective. *Exp Cell Res* **244:**372–378.
30. Nagl W. (1976) DNA endoreduplication and polyteny understood as evolutionary strategies. *Nature* **261:**614–615.
31. Brodsky WY, Uryvaeva LV. (1977) Cell polyploidy: Its relation to tissue growth and function. *Int Rev Cytol* **50:**275–332.
32. Larkins BA, Dilkes BP, Dante RA, *et al*. (2001) Investigating the hows and whys of DNA endoreduplication. *J Exp Bot* **52:**183–192.
33. Hülskamp M, Schnittger A, Folkers U. (1999) Pattern formation and cell differentiation: Trichomes in *Arabidopsis* as a genetic model system. *Int Rev Cytol* **186:**147–178.
34. Szymanski DB, Lloyd AM, Marks MD. (2000) Progress in the molecular genetic analysis of trichome initiation and morphogenesis in *Arabidopsis*. *Trends Plant Sci* **5:**214–219.
35. Folker U, Berger J, Hülskamp M. (1997) Cell morphogenesis of trichomes in *Arabidopsis*: Differential control of primary and secondary branching by branch initiation regulators and cell growth. *Development* **124:**3779–3786.
36. Kondorosi E, Roudier F, Gendreau E. (2000) Plant cell-size control: Growing by ploidy? *Curr Opin Plant Biol* **3:**488–492.
37. Galitski T, Saldanha AJ, Styles CA, *et al*. (1999) Ploidy regulation of gene expression. *Science* **285:**251–254.
38. Mizukami Y. (2001) A matter of size: Development control of organ size in plants. *Curr Opin Plant Biol* **4:**533–539.
39. Lee HC, Chiou DW, Chen WH, *et al*. (2004) Dynamics of cell growth and endoreduplication during orchid flower development. *Plant Sci* **166:**659–667.
40. Boudolf V, Vlieghe K, Beemster GTS, *et al*. (2004) The plant-specific cyclin-dependent kinase CDKB1;1 and transcription factor E2Fa-DPa control the balance of mitotically dividing and endoreduplicating cells in *Arabidopsis*. *Plant Cell* **16:**2683–2692.

41. Weinl C, Marquardt S, Kuijt SJ, *et al.* (2005) Novel functions of plant cyclin-dependent kinase inhibitors, ICK1/KRP1, can act non-cell-autonomously and inhibit entry into mitosis. *Plant Cell* **17:**1704–1722.
42. Engelen-Eigles G, Jones RJ, Phillips RL. (2000) DNA endoreduplication in maize endosperm cells: The effect of exposure to short-term high temperature. *Plant Cell Environ* **23:**657–663.
43. Cookson SJ, Radziejwoski A, Granier C. (2006) Cell and leaf size plasticity in *Arabidopsis*: What is the role of endoreduplication? *Plant Cell Environ* **29:**1273–1283.
44. Gendreau E, Traas J, Desnos T, *et al.* (1997) Cellular basis of hypocotyl growth in *Arabidopsis thaliana. Plant Physiol* **114:**295–305.
45. Gendreau E, Hofte H, Grandjean O, *et al.* (1998) Phytochrome controls the number of endoreduplication cycles in the *Arabidopsis thaliana* hypocotyl. *Plant J* **13:**221–230.
46. Ceccarelli M, Santantonio E, Marmottini F, *et al.* (2006) Chromosome endoreduplication as a factor of salt adaptation in *Sorghum bicolor. Protoplasma* **227:**113–118.

Chapter 7

The Cytogenetics of *Phalaenopsis* Orchids

Yen-Yu Kao[*,†,‡], *Chih-Chung Lin*[†], *Chien-Hao Huang*[†] *and Yi-Hsueh Li*[†]

All *Phalaenopsis* species have the same chromosome number ($2n = 2x = 38$), but their karyotypes and genome sizes vary markedly. The variation is positively correlated with the amount of constitutive heterochromatin in the genome. Genomic *in situ* and Southern hybridizations indicate that species with large genomes contain more repetitive sequences; however, species with small genomes also have their own specific repetitive sequences. One family of tandem repeats named Pvr I consisting of 7-bp repeat units was isolated and characterized. Chromosomal locations of these repeats coincide with heterochromatic blocks.

7.1 Introduction

The genus *Phalaenopsis* (Orchidaceae) represents an important source for floricultural commodity. Many species in this genus and interspecific and intergeneric hybrids are of high value due to their graceful and long-lasting flowers. *Phalaenopsis* plants are slow-growing perennials

[*]Corresponding author.
[†]*Department of Botany, National Taiwan University, Taipei 106, Taiwan.*
[‡]*Institute of Molecular and Cellular Biology, National Taiwan University, Taipei 106, Taiwan.*

that take two to three years to reach maturity. In addition, each plant produces very few roots and flowers, making cytogenetic analysis difficult. According to Sweet,[1] the genus *Phalaenopsis* is composed of approximately 47 species that are grouped into nine sections, while Christenson[2] recognizes 63 species in the genus and divides them into five subgenera and eight sections. These species are widely distributed, ranging from the Himalayas of northern India, through Southeast Asia to northern Australia.[1] Previous studies showed that all *Phalaenopsis* species possess the same chromosome number ($2n = 2x = 38$),[3,4] but their karyotypes differ markedly.[5,6] Arends[7] studied meiotic chromosome associations at metaphase I of eight interspecific hybrids. His results indicated that hybrids between species both with large chromosomes (e.g. *P. amboinensis* × *P. mannii*) and between species both with small chromosomes (e.g. *P. amabilis* × *P. stuartiana*) had complete bivalent formation, whereas hybrids between one species with large and the other species with small chromosomes (e.g. *P. mannii* × *P. equestris*) showed low frequencies of bivalents.

Two molecular techniques, fluorescence *in situ* hybridization (FISH) and genomic *in situ* hybridization (GISH), have been developed and shown as effective tools for genome research. FISH is useful for localization of specific DNA sequences on chromosomes,[8] while GISH, in which total genomic DNA from one species is the probe, can discriminate parental chromosomes in interspecific or intergeneric hybrids.[9,10,11] In addition, genomic Southern hybridization reveals the amount and distribution of repetitive DNA sequences in genomes, and is therefore also a useful tool to achieve an understanding of genome differentiation between species.[12,13] The karyotypes, genome organization and localization of tandemly repetitive sequences on chromosomes of some representative *Phalaenopsis* species were studied by us in the past few years. The results from these studies are summarized below.

7.2 Karyotypes

Karyotypes of nine *Phalaenopsis* species were analyzed based on Feulgen-stained somatic metaphase chromosomes prepared from root tips.[14] Consistent with previous studies,[4,6] all the species are diploids with a chromosome number of $2n = 38$, but their chromosomes vary in size and centromere position. The chromosomes of five species, *P. aphrodite*, *P. stuartiana*, *P. equestris*, *P. cornu-cervi* and *P. lueddemanniana*, are small and uniform in size (1–2.5 μm) and are all metacentric or submetacentric (Fig. 7.1A–B; Table 7.1). However, two species, *P. venosa* and *P. amboinensis* have strongly asymmetrical and bimodal karyotypes. They possess small, medium and large chromosomes, most of them being subtelocentric or acrocentric (Fig. 7.1C). These two species differ in the number of small chromosomes: *P. venosa* has 12 small chromosomes whereas *P. amboinensis* has 8 (Table 7.1). The karyotype of

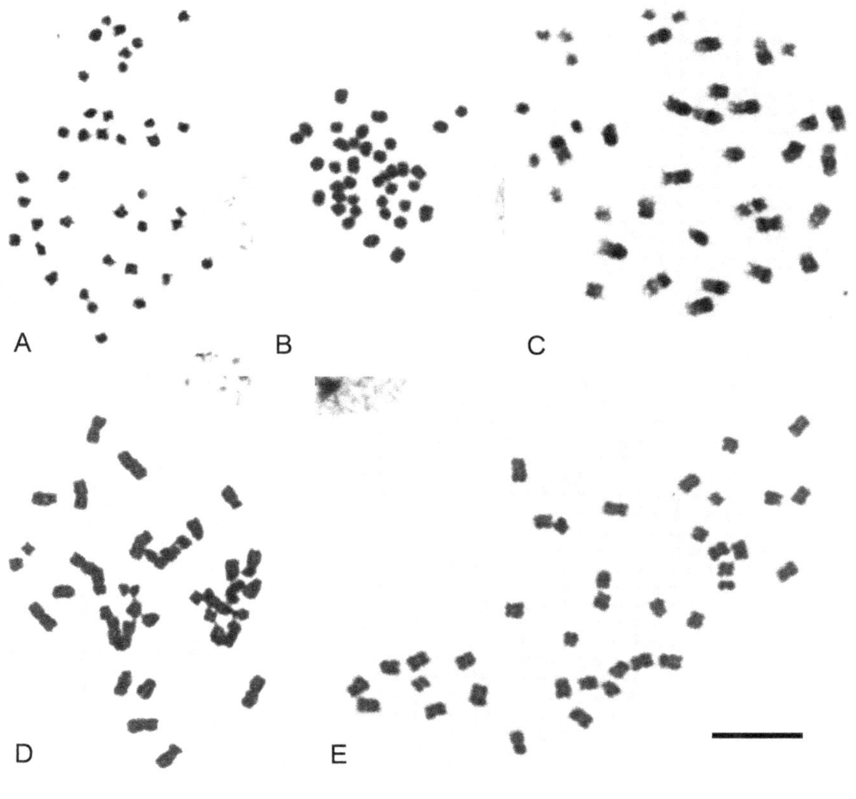

Fig. 7.1. Feulgen-stained somatic metaphase chromosomes. (**A**) *Phalaenopsis stuartiana*; (**B**) *P. cornu-cervi*; (**C**) *P. venosa*; (**D**) *P. violacea*; (**E**) *P. mannii*. Scale bar = 10 μm.

Table 7.1. Somatic Metaphase Chromosomes, Total Chromosome Volume (TCV), Nuclear DNA Content and Constitutive Heterochromatin (CH) of Nine *Phalaenopsis* Species

Taxon[a]	TSRI[b] Accession Number	2*n*	Chromosome Size (μm)[c]			TCV[c] (μm³)	Nuclear DNA Content[d] (pg 2C⁻¹)	CH[c] (% Nuclear Area)
			< 2.0	2.0–2.5	> 2.5			
Phalaenopsis								
Sect. *Phalaenopsis*								
P. aphrodite Rchb.f.	W1	38	38	0	0	ND	2.80 ± 0.06	1.87 ± 0.24
P. stuartiana Rchb.f.	W40	38	38	0	0	23.70 ± 4.51	3.13 ± 0.07	1.77 ± 0.42
Sect. *Stauroglottis*								
P. equestris (schauer) Rchb.f.	W9	38	38	0	0	22.29 ± 4.04	3.37 ± 0.05	2.74 ± 0.33
Sect. *Polychilos*								
P. cornu-cervi Rchb.f.	W8	38	34	4	0	35.91 ± 5.75	6.44 ± 0.16	2.44 ± 0.47
P. mannii Rchb.f.	W25	38	0	6	32	86.06 ± 19.06	13.50 ± 0.12	11.32 ± 1.08
Sect. *Amboinenses*								
P. amboinensis J. J. Smith	W2	38	8	4	26	ND	14.36 ± 0.19	14.17 ± 1.08
P. venosa Shim & Fowl	W42	38	12	4	22	79.76 ± 24.98	9.52 ± 0.27	13.25 ± 0.88
Sect. *Zebrinae*								
P. lueddemanniana Rchb.f.	W23	38	28	10	0	44.58 ± 3.48	6.49 ± 0.22	5.51 ± 0.46
P. violacea Witte (Bornean form)	W43	38	6	6	26	87.84 ± 8.32	15.03 ± 0.21	27.17 ± 2.03

[a]Classification of Sweet.[1]
[b]Taiwan Sugar Research Institute.
[c]Kao *et al.*[14]
[d]Lin *et al.*[17]
ND: not determined.

P. violacea is also bimodal, with12 chromosomes being small or medium and 26 chromosomes, large (Table 7.1). The 16 largest chromosomes are metacentric or submetacentric, while the remaining 10 large chromosomes are subtelocentric or acrocentric (Fig. 7.1D). *P. mannii* has comparatively more symmetrical karyotype, with 6 chromosomes being medium and 32, large (Table 7.1). Most of its chromosomes are metacentric or submetacentric (Fig. 7.1E).

The variation in genome size between related species is attributed to the difference in the amount of heterochromatin.[15,16] In order to understand the causes of karyotype variation in *Phalaenopsis* species, the amount of constitutive heterochromatin (CH) in the cell cycle from interphase was estimated and compared (Fig. 7.2; Table 7.1). The value, expressed as a percentage of the total area of CH in the interphase

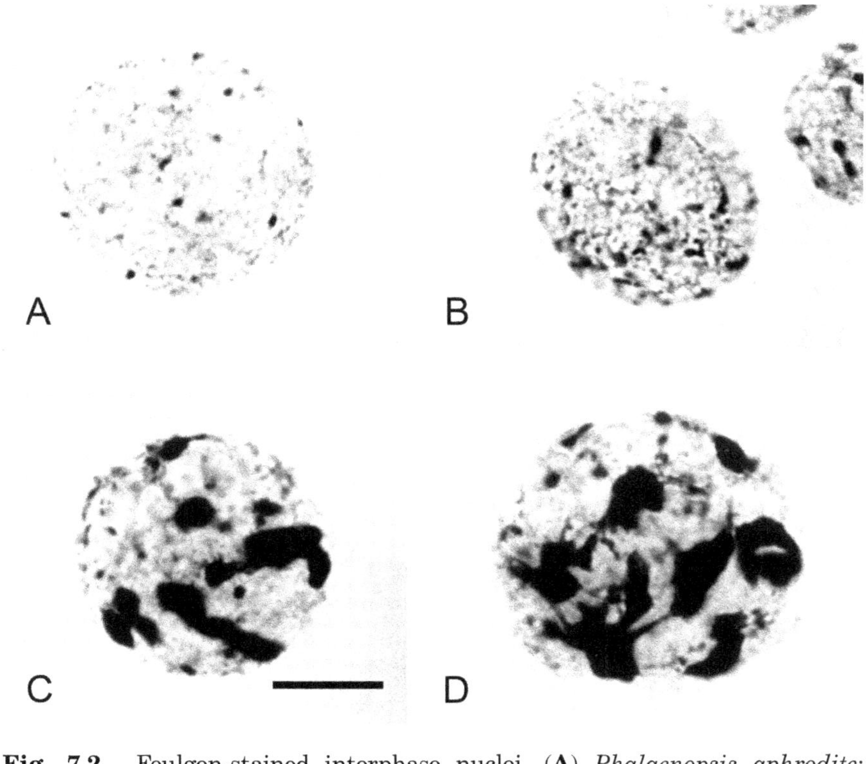

Fig. 7.2. Feulgen-stained interphase nuclei. (**A**) *Phalaenopsis aphrodite*; (**B**) *P. lueddemanniana*; (**C**) *P. amboinensis*; (**D**) *P. violacea*. Scale bar = 10 μm.

nucleus, showed a range of a 15-fold variation, ranging from 1.77% in *P. stuartiana* to 27.17% in *P. violacea* (Table 7.1). The amount of CH is positively correlated with the nuclear DNA content as reported by Lin *et al.*[17] (Table 7.1). Therefore, we suggested that differential accumulation of CH may be a major cause for karyotype variation in *Phalaenopsis* orchids.

The distribution of CH at late prophase/early metaphase chromosomes indicates that large chromosomes generally contain more CH than small ones, and this is true for chromosomes both within and between species (Fig. 7.3). In *P. aphrodite*, *P. stuartiana*, *P. equestris*,

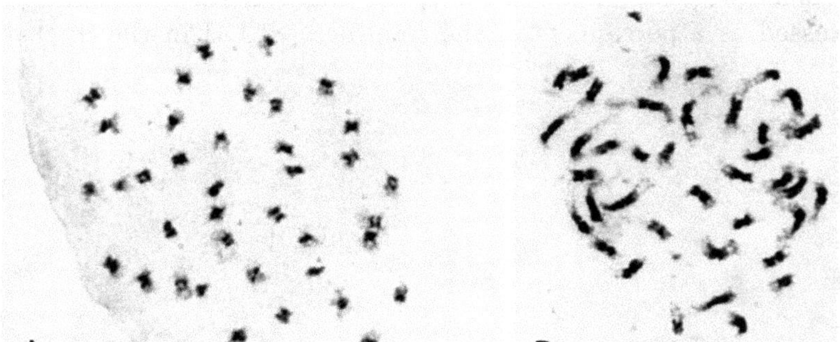

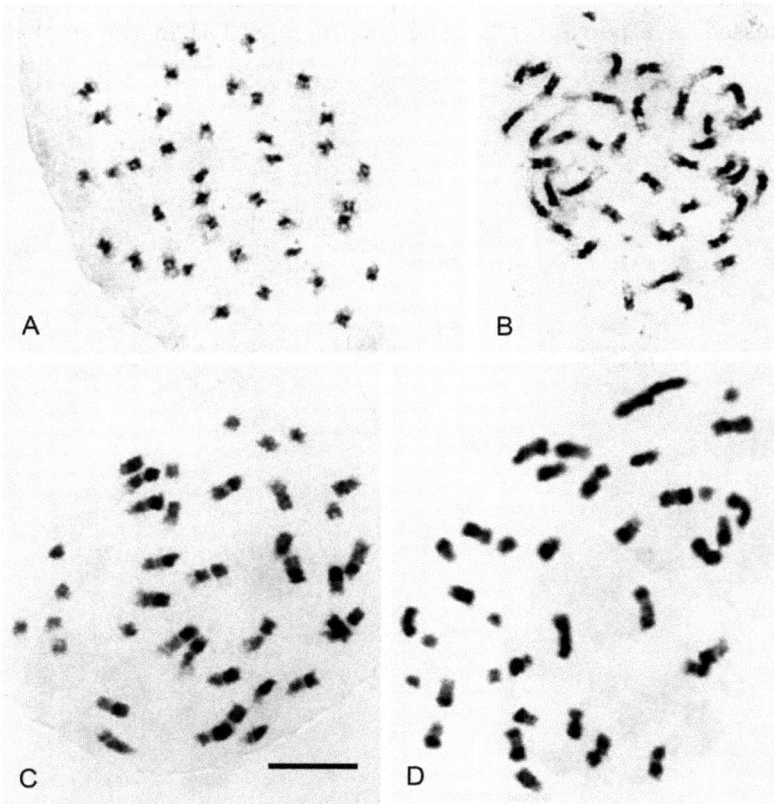

Fig. 7.3. Feulgen-stained late prophase/early metaphase chromosomes. (**A**) *Phalaenopsis aphrodite*; (**B**) *P. cornu-cervi*; (**C**) *P. venosa*. (**D**) *P. violacea*. Scale bar = 10 μm.

P. cornu-cervi and *P. lueddemanniana*, all species containing small chromosomes, CH mainly clusters around centromeres (Fig. 7.3A–B). In *P. venosa*, *P. amboinensis* and *P. violacea*, both ends of large metacentric/submetacentric chromosomes and the ends of the long arms of subtelocentric/acrocentric chromosomes possess large blocks of CH, whereas small chromosomes have little or no CH (Fig. 7.3C–D). The heterochromatin blocks in *P. mannii* are smaller in size compared with those in the species with bimodal karyotypes, and the distribution of CH is not restricted to chromosome ends. To understand the base composition of CH, root-tip cells were stained with the fluorochrome 4′-6-diamidino-2-phenyl-indole (DAPI). The result showed that DAPI-positive bands coincide with CH regions as revealed by Feulgen staining. Since DAPI binds preferentially to AT-rich regions,[18] the result suggests that the DNA of CH in *Phalaenopsis* orchids is rich in AT base pairs.

7.3 Genome Organization and Relationships Revealed by GISH

Meiotic chromosome associations in interspecific hybrids have long been used to indicate genome homology between plant species,[19] but the degree of chromosome pairing is not the only criterion for evaluating species relationships. Additionally, analysis of chromosome pairing in *Phalaenopsis* is difficult because: 1) each plant produces very few flowers, hindering collection of sufficient microsporocytes at the right stages for analysis; 2) microsporocytes are enclosed in a thick callose wall, which hampers stain penetration; and 3) meiotic chromosomes cannot be spread well due to clumping and stickiness. GISH has been shown to be a powerful tool for identifying genome constitution,[20,21] and therefore, may be a method complementing analysis of meiotic chromosome pairing. In order to understand the relationships of *Phalaenopsis* species with various sizes of genomes, we examined seven interspecific hybrids of *Phalaenopsis* by GISH.[22] The results are listed in Table 7.2. In hybrid *P. aphrodite* × *P. sanderiana*, in which both parents contained small genomes, the parental chromosome sets could not be distinguished even in the presence of 100-fold blocking DNA (Fig. 7.4A–B). A similar result was observed in hybrid *P. mannii* × *P. violacea*, in which both parents possessed large genomes (Fig. 7.4C–D). Normal chromosome pairing in hybrids with

Table 7.2. The Results from GISH in Seven *Phalaenopsis* Interspecific Hybrids

Hybrid	2*n*	Probe DNA	Differentiation of Parental Chromosomes[a]
Hybrid between species with small genomes:			
P. aphrodite × *P. sanderiana*	38	*P. sanderiana*	–
Hybrid between species with large genomes:			
P. mannii × *P. violacea*	38	*P. violacea*	–
P. mannii × *P. violacea*	38	*P. mannii*	–
Hybrids between species with small and large genomes:			
P. amboinensis × *P. stuartiana*	38	*P. amboinensis*	+
P. amboinensis × *P. stuartiana*	38	*P. stuartiana*	+
P. violacea × *P. equestris*	38	*P. violacea*	+
P. violacea × *P. equestris*	38	*P. equestris*	+
P. mannii × *P. stuartiana*	38	*P. mannii*	+
P. mannii × *P. stuartiana*	38	*P. stuartiana*	+
P. mannii × *P. aphrodite*	38	*P. mannii*	+
P. mannii × *P. aphrodite*	38	*P. aphrodite*	+
Hybrid between species with medium and large genomes:			
P. mannii × *P. lueddemanniana*	38	*P. mannii*	±

a: No differentiation; ±: slight differentiation; +: clear differentiation.

similar sizes of genomes was reported by Arends.[7] A comparison of our GISH results with those of Arends supports the suggestion that species with similar genome sizes have close phylogenetic relationships.

In hybrids *P. mannii* × *P. stuartiana*, *P. mannii* × *P. aphrodite*, *P. amboinensis* × *P. stuartiana* and *P. violacea* × *P. equestris*, the genomes of *P. mannii*, *P. amboinensis* and *P. violacea* are much larger than those of *P. stuartiana*, *P. aphrodite* and *P. equestris*,[14,17] and the parental species of these hybrids belong to different sections.[1,2] In general, results obtained from these hybrids were similar (Table 7.2). When genomic DNAs from species with large genomes were used as probes, the parental chromosome sets could be distinguished in all

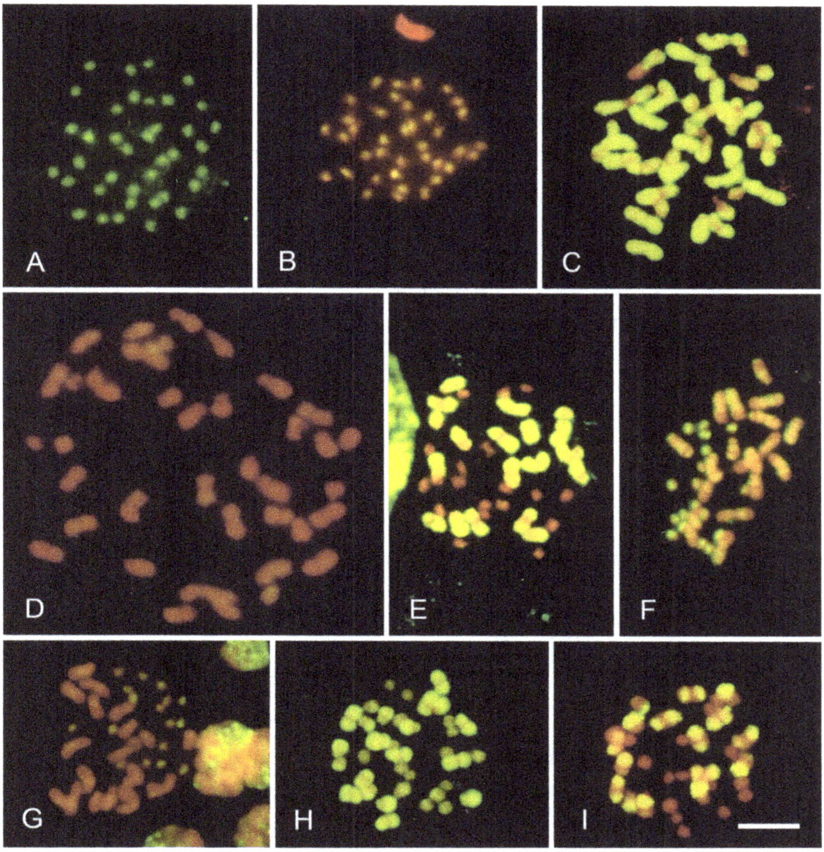

Fig. 7.4. Genomic *in situ* hybridization in *Phalaenopsis* interspecific hybrids. Somatic metaphase chromosomes prepared from the hybrids were probed with FITC-labeled genomic DNA from one parent in the presence or absence of unlabeled (blocking) DNA from the other parent. Chromosomes were counterstained with propidium iodide. (**A–B**) Hybrid *P. aphrodite* × *P. sanderiana* chromosomes probed with *P. sanderiana* DNA in the absence (**A**) and presence of 100× blocking DNA (**B**); (**C–D**), Hybrid *P. mannii* × *P. violacea* chromosomes probed with *P. mannii* DNA in the absence (**C**) and presence of 100× blocking DNA (**D**); (**E**), Hybrid *P. mannii* × *P. stuartiana* chromosomes probed with *P. mannii* DNA in the absence of blocking DNA; (**F–G**) Hybrid P. *mannii* × *P. stuartiana* chromosomes probed with *P. stuartiana* DNA in the absence (**F**) and presence of 50× blocking DNA (**G**); (**H–I**), Hybrid *P. mannii* × *P. luedde-manniana* chromosomes probed with *P. mannii* DNA in the absence (**H**) and presence of 50× blocking DNA (**I**). Scale bar = 10 μm.

hybrids without the application of blocking DNA (Fig. 7.4E). The strength and distribution of hybridization signals in these hybrids indicate that species with large genomes contain more repetitive sequences than species with small genomes and confirm distant relationships between the two parental species based on morphological characters.[1,2] The results are consistent with Arend's findings that hybrids between species with dissimilar sizes of genomes showed poor chromosome pairing.[7] When total DNA from species with small genomes was used as the probe, in the absence of blocking DNA, strong signals dispersed on chromosomes of the species with small genomes (Fig. 7.4F). After addition of 50× unlabeled blocking DNA from species with large genomes, hybridization signals were restricted to centromeric regions of small chromosomes (Fig. 7.4G). These results suggest that species with small genomes have their own specific sequences at or around centromeres.

In hybrid *P. mannii* × *P. lueddemanniana*, nuclear DNA contents of these two parents differ approximately twofold; addition of 50-fold blocking DNA was necessary to distinguish the parental chromosomes (Fig. 7.4H–I). This observation indicates that *P. mannii* and *P. lueddemanniana* share a high degree of sequence homology and, therefore, have a close phylogenetic relationship. This is consistent with the cytological work of Arends,[7] who found complete bivalent formation at the first meiotic division in hybrid *P. lueddemanniana* × *P. mannii*.

7.4 Repetitive Sequences

7.4.1 *(GA)$_n$* microsatellites

Microsatellites, also called simple sequence repeats (SSRs), are tandem repeated arrays of short core sequence. One microsatellite, $(GA)_{11}$, has been mapped physically to seven *Phalaenopsis* species and *Doritis pulcherrima*.[23] $(GA)_{11}$ signals are dispersed on the chromosomes of *P. amboinensis* and *P. violacea*, but clustered in the centromeric regions of *P. stuartiana*, *P. equestris*, *P. aphrodite*, *P. mannii*, *P. lueddemanniana* and *D. pulcherrima* chromosomes (Fig. 7.5A–B). Signals were much stronger in *P. stuartiana* and *P. aphrodite* than in other species. In GISH experiment of hybrids between species with small and large

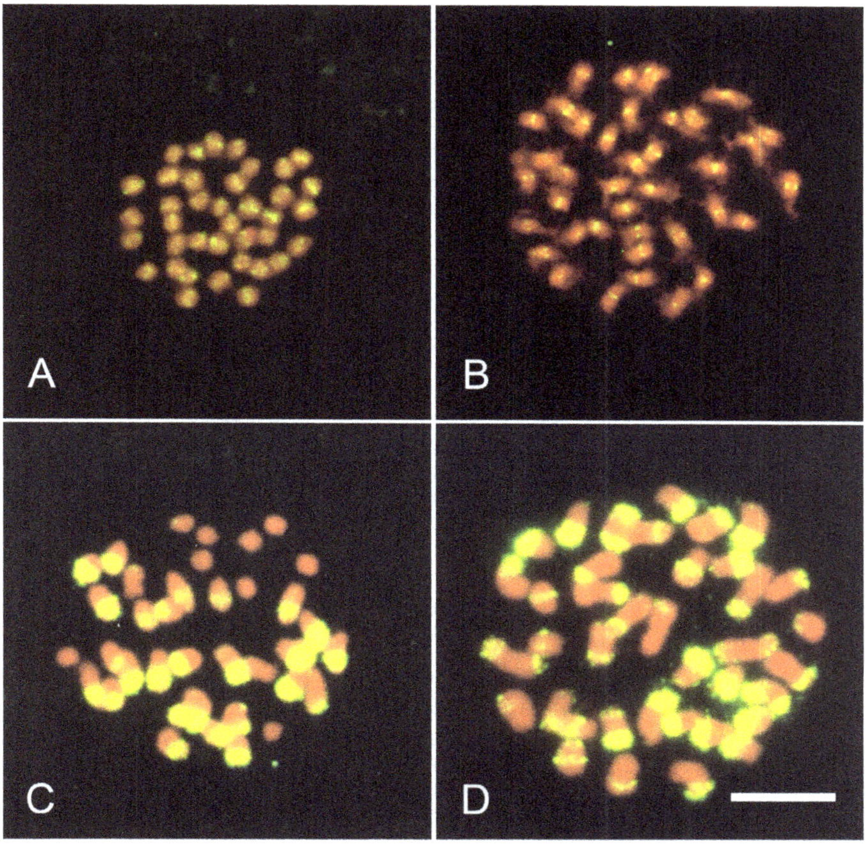

Fig. 7.5. Fluorescence *in situ* hybridization of $(GA)_{11}$ microsatellite (**A–B**) and Pvr-93 sequences (**C–D**) on chromosomes of *Phalaenopsis* species. The probe was labeled with digoxigenin and detected with anti-digoxigenin-fluorescein. Chromosomes were counterstained with propidium iodide. (**A**) *Phalaenopsis stuartiana*; (**B**) *P. equestris*; (**C**) *P. venosa*; (**D**) *P. mannii*. Scale bar = 10 μm.

genomes mentioned above, when total DNA from species with small genomes was used as the probe in the presence of 50× blocking DNA, hybridization signals were restricted to the centromeric regions of the small chromosomes (Fig. 7.4G). From these, we suspect that GISH signals in the centromeric regions of *P. stuartiana*, *P. aphrodite*, and *P. equestris* chromosomes were from hybridization of $(GA)_n$ microsatellite sequences.

7.4.2 *Pvr I tandem repeats*

Genomic Southern hybridization was used to investigate the amount and distribution of repetitive DNA sequences in *Phalaenopsis*.[24] The hybridization patterns in *P. violacea* and *P. mannii* were similar when the total genomic DNA from *P. violacea* was the probe. This result suggests that the kind and amount of repetitive sequences in these two species are similar. This may explain the difficulty of discriminating the chromosomes of these two species by GISH (Fig. 7.4C–D; Table 7.2). In addition, the probe from *P. violacea* hybridized strongly to DNAs from species with large genomes, but weakly to DNAs from species with small genomes. However, when total genomic DNA from *P. aphrodite* was used as a probe, an opposite result was obtained, i.e. species with small genomes showed stronger signals than did species with large genomes. These results are consistent with the observations of GISH that species with large genomes contain more repetitive sequences and species with small genomes have their own specific sequences.

One family of tandemly repeated sequences, Pvr I, was isolated from *P. violacea*.[24] Twelve clones of this family were sequenced. Sequences of these clones are not found in the EMBL/Genbank database. All sequences are AT-rich at the 5′- and 3′-ends and consist of 7-bp repeat units organized in tandems, ranging from 13 to 22 copies. The consensus sequence of these repeats is GTAAGCC. Physical mapping of one sequence of this family, Pvr-93, was preformed by FISH on mitotic metaphase chromosomes prepared from the root tips of *Phalaenopsis* species and the related species *Doritis pulcherrima*. The results revealed that in species with large genomes, including *P. violacea*, *P. mannii*, *P. amboinensis*, *P. venosa* and *D. pulcherrima*, Pvr-93 sequences were located in the constitutive heterochromatin regions of large chromosomes corresponding to DAPI-positive bands (Fig. 7.5C–D). No hybridization signals were observed on the small chromosomes of *P. aphrodite* and *P. equestris*. These results indicate that Pvr I sequences are species-specific and reside in AT-rich heterochromatic blocks of large chromosomes.

7.5 Conclusion

The nine representative *Phalaenopsis* species studied by us have the same chromosome number ($2n = 2x = 38$), but their karyotypes differ markedly. We suggest that differential accumulation of CH is a major cause for karyotype variation. Our results from FISH, GISH and genomic Southern hybridization demonstrate that the *Phalaenopsis* species possess different kinds and amounts of repetitive sequences. A more detailed understanding of genome organization and relationships in *Phalaenopsis* will result from comparative studies of chromosome pairing at meiosis, isolation and characterization of species-specific sequences.

Acknowledgment

We are grateful to Dr. Chi-Chang Chen for his critical reading of this manuscript.

References

1. Sweet HR. (1980) *The Genus Phalaenopsis*. Day Printing Corporation, California.
2. Christenson EA. (2001) *Phalaenopsis: A Monograph*. Timber Press, Portland, Oregon.
3. Sagawa Y. (1962) Cytological studies of the genus *Phalaenopsis*. *Amer Orchid Soc Bull* **31:**459–465.
4. Sagawa Y, Shoji T. (1968) Chromosome numbers in *Phalaenopsis* and *Doritis*. *Bul Pacific Orchid Soc* **26:**9–13.
5. Woodard JW. (1951) Some chromosome numbers in *Phalaenopsis*. *Amer Orchid Soc Bull* **20:**356–358.
6. Shindo K, Kamemoto H. (1963) Karyotype analysis of some species of *Phalaenopsis*. *Cytologia* **28:**390–398.
7. Arends JC. (1970) Cytological observations on genome homology in eight interspecific hybrids of *Phalaenopsis*. *Genetica* **41:**88–100.
8. Jiang J, Gill BS. (1994) Nonisotopic in situ hybridization and plant genome mapping: the first 10 years. *Genome* **37:**717–725.
9. Schwarzacher T, Leitch AR, Bennett MD, Heslop-Harrison JS. (1989) *In situ* localization of parental genomes in a wide hybrid. *Ann Bot* **64:**315–324.
10. Ørgaard M, Heslop-Harrison JS. (1994) Investigations of genome relationships between *Leymus*, *Psathyrostachys*, and *Hordeum* inferred by genomic DNA: DNA *in situ* hybridization. *Ann Bot* **73:**195–203.
11. D'Hont A, Paget-Goy A, Escoute J, Carreel F. (2000) The interspecific genome structure of cultivated banana, *Musa spp.* revealed by genomic DNA *in situ* hybridization. *Theor Appl Genet* **100:**177–183.

12. Ørgaard M, Heslop-Harrison JS. (1994) Relationships between species of *Leymus*, *Psathyrostachys*, and *Hordeum* (*Poaceae*, *Triticeae*) inferred from Southern hybridization of genomic and cloned DNA probes. *Plant Syst Evol* **189:**217–223.
13. Aggarwal RK, Brar DS, Khush GS. (1997) Two new genomes in the *Oryza* complex identified on the basis of molecular divergence analysis using total genomic DNA hybridization. *Mol Gen Genet* **254:**1–12.
14. Kao YY, Chang SB, Lin TY, *et al.* (2001) Differential accumulation of heterochromatin as a cause for karyotype variation in *Phalaenopsis* orchids. *Ann Bot* **87:**387–395.
15. Flavell R. (1982) Sequence amplification, deletion and rearrangement: Major sources of variation during species divergence. In: Dover FA, Flavell RB (eds.), *Genome Evolution*. Academic Press, London, pp. 301–323.
16. Kubis S, Schmidt T, Heslop-Harrison JS. (1998) Repetitive DNA elements as a major component of plant genomes. *Ann Bot* **82** (**Suppl. A**)**:**45–55.
17. Lin S, Lee HC, Chen WH, *et al.* (2001) Nuclear DNA contents of *Phalaenopsis* species and *Doritis pulcherrima*. *J Amer Soc Hort Sci* **126:**195–199.
18. Sumner AT. (1990) *Chromosome Banding*. Unwin Hyman, London.
19. Singh RJ. (2003) *Plant Cytogenetics*. 2nd Edition. CRC Press, Boca Raton, Florida, pp. 277–306.
20. Bennett ST, Kenton AY, Bennett MD. (1992) Genomic *in situ* hybridization reveals the allopolyploid nature of *Milium montianum* (Gramineae). *Chromosoma* **101:**420–424.
21. Fukui K, Shishido R, Kinoshita T. (1997) Identification of the rice D-genome chromosomes by genomic *in situ* hybridization. *Theor Appl Genet* **95:** 1239–1245.
22. Lin CC, Chen YH, Chen WH, *et al.* (2005) Genome organization and relationships of *Phalaenopsis* orchids inferred from genomic in situ hybridization. *Bot Bull Acad Sin* **46:**339–345.
23. Li YH. (2000) Investigation on the physical distribution of microsatellites in *Phalaenopsis* orchids by fluorescence *in situ* hybridization. Master thesis, Department of Botany, National Taiwan University, Taipei, Taiwan.
24. Huang CH. (1999) Isolation and characterization of two repetitive DNA sequences in *Phalaenopsis* orchids. Master thesis, Department of Botany, National Taiwan University, Taipei, Taiwan.

Chapter 8

Analysis of the Chloroplast Genome of *Phalaenopsis aphrodite*

Ching-Chun Chang[*,†], *Hsien-Chia Lin*[†] *and Wun-Hong Zeng*[†]

The complete nucleotide sequence of the chloroplast genome of the Taiwan moth orchid (*Phalaenopsis aphrodite* subsp. *formosana*) was determined. The circular, double-stranded DNA of 148,964 bp comprises a pair of inverted repeats of 25,732 bp, which are separated by a small single copy (SSC) and a large single copy (LSC) region of 11,543 and 85,957 bp, respectively. The genome contains 76 protein coding genes, four ribosomal RNA genes, 30 tRNA genes, and 24 putative open reading frames (ORFs). Seventeen genes are intron-containing, including 6 tRNA and 11 protein coding genes. Unlike other chloroplast genomes of photosynthetic angiosperms, which have a complete set of genes in the 11 subunits of NADH dehydrogenase, the chloroplast genome of *Phalaenopsis* completely lacks the *ndhA*, *ndhH*, and *ndhF* genes. The other eight *ndh* genes have various degrees of nucleotide insertion/deletion, and they are all frameshifted. This loss results in the SSC region in *Phalaenopsis* being the shortest among known photosynthetic angiosperms.

8.1 Introduction

Chloroplasts, in which photosynthesis takes place, have distinct functional genomes. They were once free-living cyanobacteria and even now contain preserved remnants of eubacterial genomes,[1] but with some of their genes

*Corresponding author.

†*Institute of Biotechnology, National Cheng Kung University, Tainan, Taiwan.*

having been transferred to the nucleus of the cell during the course of evolution.[2,3] Since the complete plastid genome sequences of tobacco and liverwort were reported in 1986,[4,5] over 30 complete sequences representing the major lineages of vascular plants have been made available in the genomic database of NCBI (http://www.ncbi.nlm.nih.gov/genomes/ORGANELLES/plastids_tax.html). The DNA molecules in the chloroplast genome of photosynthetic plants are circular and range in size from 116 to 204 kb. With few exceptions, such as in conifers[6] and *Medicago* (accession number NC_003119, unpublished), most of the genomes that have been examined have a large (5–76 kb) inverted repeat (IR) segment, which accounts for the variation in length of the genomes.[2,7,6] Two segments of the IRs are divided by a large single copy (LSC) and a small single copy (SSC) region.[2,7] Among vascular plants, chloroplast genome contains 110–130 genes, and most of these genes code for proteins mostly involved in photosynthesis or gene expression, with the remainder being transfer RNA or ribosomal RNA genes.[8] Furthermore, the polycistronic transcription units of the known chloroplast genomes are also highly conserved.[7]

Although the plastid genomes of vascular plants are conserved in the overall structure,[7] comparative genomic studies have recently revealed many evolutionary hotspots and single nucleotide polymorphism that are useful for phylogenetic studies below the family and subspecies level.[9–12] Furthermore, genome-wide analysis of transcripts in the chloroplast has revealed that plants have undergone dramatic changes in both the levels and patterns of editing, from hornworts (1.2%) and ferns (0.38%) to seed plants (less than 0.05%).[13,14] In addition to phylogenetic studies, complete genome sequences provide useful information for designing expression cassettes for chloroplast genetic manipulation and for elucidating functional genomics.[15–17]

With approximately 30,000 species, orchids are one of the largest and most diverse families of flowering plants.[18] The mode of chloroplasts DNA transmission in orchids is uniparentally inherited from female ancestors.[19,20] Based on molecular markers from plastid genes, the phylogenetic relationship has been investigated at the family and subfamily level within Orchidaceae.[21,22] As a part of a larger effort to preserve the native species and to utilize its chloroplast genomes for various biotechnological purposes, we sequenced the complete chloroplast genome of *Phalaenopsis aphrodite* subsp. *formosana*. It is anticipated that knowledge of the chloroplast genome of this species will be useful for commercial breeding and for phylogenetic classification within Orchidaceae as well as for genome-based analysis in angiosperm evolution.

8.2 *Phalaenopsis* Chloroplast DNA Extraction, Genome Sequencing, and Gene Annotation

Leaves of *Phalaenopsis aphrodite* subsp. *formosana* were obtained from seedlings at the four-leaf stage of development. Intact chloroplasts were fractionated with step percoll (40–80%) gradient.[23] The chloroplast DNA was further isolated according to a CTAB-based protocol.[24] Chloroplast DNA was randomly fragmented into 2 to 3 kb pieces through hydroshear with GeneMachine (Genomic Solutions Inc., Michigan, USA), and then cloned into the pBluescriptSK vector. After transformation, positive shotgun clones were propagated on microtiter plates. Plasmid DNAs were isolated and used as templates for sequence analysis.

The sequencing reaction was performed using the BigDye terminator cycle sequencing kit (Applied Biosystems, USA), according to the protocol recommended by the manufacturer. The DNA sequencer used was an Applied Biosystems ABI 3700. The sequence data from both ends of each shotgun clone were accumulated, trimmed, aligned, and assembled using Phred-Phrap programs (Phil Green, University of Washington, Seattle, USA). The sequence data were accumulated to 1,728,000 nucleotides, which is 11.6X coverage for the chloroplast genome of *P. aphrodite* subsp. *formosana*. Database searches were conducted with the BLAST algorithm provided by the National Center for Biotechnology Information (NCBI). The tRNAscan program was used to assign the tRNA genes.

8.3 *Phalaenopsis* Chloroplast Genome Properties

8.3.1 *Overall genome structure*

The complete circular chloroplast DNA of *Phalaenopsis aphrodite* subsp. *formosana* (the Taiwan moth orchid) is 148,964 bp (accession number AY916449) long and can be separated into two regions by a pair of inverted repeat regions (IR) of 25,732 bp (Fig. 8.1). The large single copy region (LSC) and the small single copy region (SSC) are 85,957 bp and 11,543 bp long, respectively. The overall A+T content of the chloroplast genome of *P. aphrodite* subsp. *formosana* is 63.4%, which is close to the A+T percentages of rice (61%), maize (61.5%), wheat (61.7%), wild rice (61%), and sugarcane (61.6%). The A+T contents in the LSC and SSC region are higher, being 66.1% and 71.9%, respectively, than in the IR regions (56.8%). The gene order in the *Phalaenopsis* chloroplast

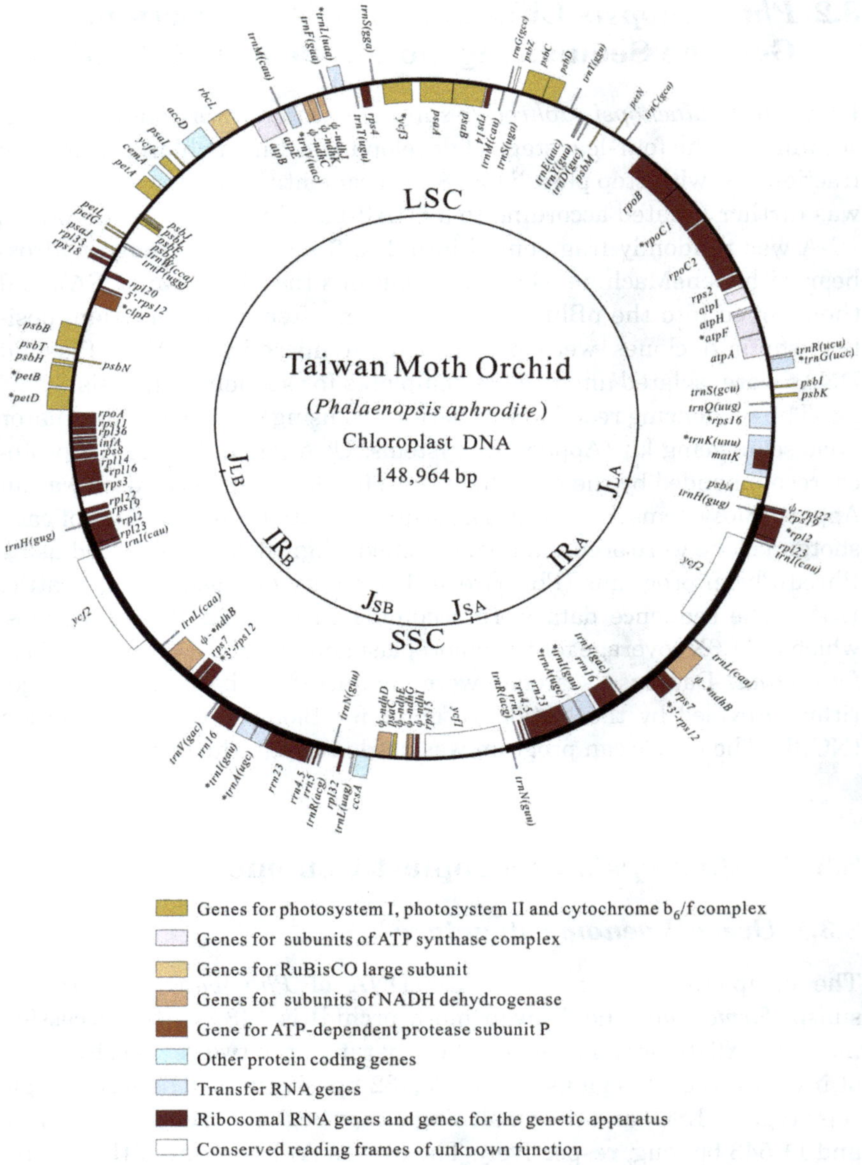

Fig. 8.1. Gene map of *Phalaenopsis aphrodite* subsp. *formosana* chloroplast genome.[30] Genes shown inside circle are transcribed clockwise; those outside are transcribed counterclockwise. Genes belonging to different functional groups are color coded. Asterisks indicate genes containing introns.

genome is more similar to that of the known dicots than to the grasses, as in the latter, gene rearrangements caused by inversion are extensive in the LSC region.[9,25–28]

8.3.2 *Gene content, intron, and codon usage*

Table 1 presents the content of genes in the *Phalaenopsis* chloroplast DNA. A total of 110 genes (not including genes duplicated in the IR) were identified as being scattered around the chloroplast genome. They include 76 protein coding genes, four ribosomal RNA genes, and 30 transfer RNA genes. Among them, 17 are intron-containing genes, including 6 tRNA and 11 protein coding genes (Table 2). Three of the latter genes, i.e. *clpP*, *rps12*, and *ycf3*, contain two introns. In the five grasses, there are 16 intron-containing genes, and the *clpP* and *rpoC1* genes have lost introns.[9,25–28] The *rps12* gene is a divided gene whose 5′-end exon is located in the LSC region, but whose second and third exons are in the IR regions. It appears that a *trans*-splicing mechanism is required between exon I and exon II to yield mature *rps12* transcripts. The intron of the *trnL*-UAA gene contains minisatellite repeat elements in the marsh orchid, *Orchis palustris*,[29] but in *P. aphrodite* subsp. *formosana* it does not.

In addition, 24 ORFs were identified with a threshold of 225 bp. An ORF, *orf91*, located in the complementary strand of the *rrn23* gene, was reported for the first time.[30] Its amino acid sequence is 81% identical to a hypothetical protein (NCBI accession no. ZP_00203428) encoded in the genome of *Anabaena variabilis*. We also observed the *orf91* transcripts in the total RNAs of *P. aphrodite* and tobacco by RT-PCR assay and RNA filter hybridization,[30] which demonstrated that *orf91* was actually transcribed in the chloroplasts. However, further studies are required to determine if the *orf91* transcripts are translated and to characterize the function of the translated protein.

The codon usage in *Phalaenopsis* chloroplast genome and the anticodons of 30 tRNA species are summarized in Table 3. The codon usage determined from 67 protein coding genes, is strongly biased toward A or T due to the high A-T content at the third codon positions. For example, degenerated codons at the third position exhibit usage frequencies of 65% to 81% for both A or T. The frequencies of the stop codon usage for UAA, UAG, and UGA are 48%, 27%, and 25%, respectively. These are also strongly biased toward A or T at both the second and third positions. The strong AT bias of codon usage in *Phalaenopsis* is consistent

Table 8.1. List of Gene Contents in *Phalaenopsis* Chloroplast Genome (Total 110 Genes)

Group of Genes	
RNA genes	
rRNA genes	*rrn4.5*,[b] *rrn5*,[b] *rrn16*,[b] *rrn23*[b]
tRNA genes	*trnA* (ugc),[a,b] *trnC* (gca), *trnD* (guc), *trnE* (uuc), *trnF* (gaa), *trnG* (gcc), *trnG* (ucc),[a] *trnH* (gug),[b] *trnI* (cau), *trnI* (gau),[a,b] *trnK* (uuu),[a] *trnL* (caa),[b] *trnL* (uag), *trnL* (uaa),[a] *trnM* (cau),[b] *trnfM* (cau), *trnN* (guu),[b] *trnP* (ugg), *trnQ* (uug), *trnR* (ucu), *trnR* (acg),[b] *trnS* (gcu), *trnS* (uga), *trnS* (gga), *trnT* (ggu), *trnT* (ugu), *trnV* (gac),[b] *trnV* (uac),[a] *trnW* (cca), *trnY* (gua)
Protein coding genes	
Ribosomal proteins	
Small subunit	*rps2*, *rps3*, *rps4*, *rps7*,[b] *rps8*, *rps11*, *rps12*,[a,b] *rps14*, *rps15*, *rps16*,[a] *rps18*, *rps19*[b]
Large subunit	*rpl2*,[a,b,c] *rpl14*, *rpl16*,[a] *rpl20*, *rpl22*, *rpl23*,[b] *rpl32*, *rpl33*, *rpl36*
Transcription/translation apparatus	
DNA dependent RNA polymerase	*rpoA*, *rpoB*, *rpoC1*,[a] *rpoC2*
Translation initiation factor	*infA*
Photosynthesis	
Subunits of Photosystem I	*psaA*, *psaB*, *psaC*, *psaI*, *psaJ*, *ycf3*,[a] *ycf4*
Subunits of Photosystem II	*psbA*, *psbB*, *psbC*, *psbD*, *psbE*, *psbF*, *psbH*, *psbI*, *psbJ*, *psbK*, *psbL*, *psbM*, *psbN*, *psbT*, *psbZ*
Subunits of Cytochrome b_6/f complex	*petA*, *petB*,[a] *petD*,[a] *petG*, *petL*, *petN*
Subunits of ATP synthase	*atpA*, *atpB*, *atpE*, *atpF*,[a] *atpH*, *atpI*
Subunits of NADH-dehydrogenase	*ψ-ndhB*,[a,b] *ψ-ndhC*, *ψ-ndhD*, *ψ-ndhE*, *ψ-ndhG*, *ψ-ndhI*, *ψ-ndhJ*, *ψ-ndhK*
Large subunit of RubisCO	*rbcL*

(*Continued*)

Table 8.1. *(Continued)*

Group of Genes				
Miscellaneous proteins	*accD, clpP,*[a] *matK, cemA, ccsA*			
Conserved proteins	*ycf1, ycf2*[b]			
Open reading frames	ORF114 (175_d)	ORF80 (1754_d)	ORF98 (22677_d)	ORF87B (23929_d)
	ORF103 (51817_d)	ORF88 (61993_e)	ORF87C (64106_d)	ORF85 (69399_e)
	ORF79 (70125_d)	ORF99 (70354_e)	ORF81C (74208_e)	ORF86B (86687_d)[b]
	ORF90 (93146_d)[b]	ORF81B (97939_d)[b]	ORF79 (99557_d)[b]	ORF77A (100484_d)[b]
	ORF81A (102460_d)[b]	ORF115 (102606_d)[b]	ORF86A (104928_d)[b]	ORF170 (105185_d)[b]
	ORF91 (109119_d)[b]	ORF131 (110120_d)[b]	ORF77B (112628_e)	ORF87A (122213_d)

[a]Intron containing gene.
[b]Two copies due to inverted repeat.
[c]Initiation codon created by RNA editing.
[d]ORF translated clockwise.
[e]ORF translated counterclockwise.
The position of start codon of putative ORF is given in parenthesis.

Table 8.2. The Splitting Genes on *Phalaenopsis* Chloroplast Genome

Gene	Location	Exon I	Intron I	Exon II	Intron II	Exon III
atpF	LSC	145	986	410		
clpP	LSC	71	981	292	689	252
ndhB	IR	733	532	763		
petB	LSC	6	726	642		
petD	LSC	8	865	484		
rpl2	IR	391	664	431		
rpl16	LSC	9	1251	399		
rpoC1	LSC	453	758	1608		
*rps12**	IR	114	—	232	547	26
rps16	LSC	245	914	40		
trnA-ugc	IR	38	802	35		
trnG-ucc	LSC	23	687	48		
trnI-gau	IR	37	944	35		
trnK-uuu	LSC	37	2896	35		
trnL-uaa	LSC	35	717	50		
trnV-uac	LSC	39	582	35		
ycf3	LSC	124	731	230	750	153

*The *rps12* gene is divided. The 5′-*rps12* locates on the LSC region, while the 3′-*rps12* locates on the IR region. The number indicates the length of the intron or exon in the base pairs.

with the results published elsewhere concerning the chloroplast genome of many higher plants.[5,11,25,27]

8.3.3 *Loss of ndh genes*

Ndh genes that encode the subunits of the NADH dehydrogenase complex are involved in the cyclic electron flow of photosystem I and chlororespiration in tobacco.[15] All 11 subunits of *ndh* genes are present in the chloroplast genomes of photosynthetic angiosperms so far sequenced. In contrast, in the chloroplast genomes of black pine, the *ndhA*, *ndhG*, *ndhF*, and *ndhJ* genes were completely absent, and sequences of the other seven *ndh* genes were truncated as obvious pseudogenes.[31] In *Phalaenopsis*, the *ndhA*, *ndhF*, and *ndhH* genes are completely absent from the chloroplast genome, and the other remaining eight *ndh* genes are frameshifted. The *ndhB*, *ndhC*, and *ndhJ* genes have small region of nucleotide insertion/deletion, and the *ndhD*, *ndhE*, *ndhG*, *ndhI,* and

Table 8.3. Codon Usage of the *Phalaenopsis* Chloroplast Genome

UUU	F	612	*trnF*-GAA	UCU	S	398	*trnS*-GGA	UAU	Y	475	*trnY*-GUA	UGU	C	147	*trnC*-GCA
UUC	F	331		UCC	S	215		UAC	Y	116		UGC	C	48	
UUA	L	557	*trnL*-UAA	UCA	S	254	*trnS*-UGA	UAA	—	32		UGA	—	17	*trnW*-CCA
UUG	L	386	*trnL*-CAA	UCG	S	100		UAG	—	18		UGG	W	308	
CUU	L	363	*trnL*-UAG	CCU	P	287	*trnP*-UGG	CAU	H	355	*trnH*-GUG	CGU	R	272	*trnR*-ACG
CUC	L	113		CCC	P	167		CAC	H	95		CGC	R	60	
CUA	L	226		CCA	P	200		CAA	Q	501	*trnQ*-UUG	CGA	R	250	
CUG	L	119		CCG	P	68		CAG	Q	158		CGG	R	64	
AUU	I	727	*trnI*-GAU	ACU	T	379	*trnT*-GGU	AAU	N	612	*trnN*-GUU	AGU	S	289	*trnS*-CCU
AUC	I	294		ACC	T	152		AAC	N	165		AGC	S	62	
AUA	I	423	*trnI*-CAU	ACA	T	263	*trnT*-UGU	AAA	K	663	*trnK*-UUU	AGA	R	331	*trnR*-UCU
AUG	M	395	*trnM*-CAU	ACG	T	98		AAG	K	243		AGG	R	119	
			trnfM-CAU												
GUU	V	369	*trnV*-GAC	GCU	A	473	*trnA*-UGC	GAU	D	570	*trnD*-GUC	GGU	G	433	*trnG*-GCC
GUC	V	129		GCC	A	138		GAC	D	124		GGC	G	122	
GUA	V	357	*trnV*-UAC	GCA	A	312		GAA	E	701	*trnE*-UUC	GGA	G	490	*trnG*-UCC
GUG	V	155		GCG	A	90		GAG	E	228		GGG	G	207	

ndhK genes are truncated with large deletion of 26%, 60%, 45%, 46%, and 32% of the sequences, respectively, as compared with that of tobacco. Probably, all the resident *ndh* genes were pseudogenes. The loss of functional *ndh* genes explains well why *P. aphrodite* subsp. *formosana* has the smallest SSC region among the sequenced monocots. Although the majority of the *ndh* genes have been lost from the chloroplast genome of *Pinus* and *Phalaenopsis*, the *ndhF* gene was present in all vascular plant divisions,[32] suggesting that the lost *ndhF* and other *ndh* genes in the plastid genome of *Phalaenopsis* might have been transferred to the nuclear genome. Indeed, we were able to detect the presence of *ndhA*, *ndhF*, and *ndhH* gene fragments by RT-PCR (unpublished data). Further studies will unravel the details of the lost *ndh* genes in the plastid of *P. aphrodite*.

8.3.4 *RNA editing*

RNA editing plays an important role in the regulation of gene expression in chloroplasts. The plastid *rpl2* gene of *Phalaenopsis* has an ACG codon rather than an ATG codon at the translation initiation site (positions 88242–88244) as was observed in maize.[33] The ACG rather than an ATG was also found at the initiation codon of the *ndhD* genes (positions 115315–115317), as in tobacco and spinach.[34] Furthermore, frameshifts were observed in the *ndh* genes, suggesting RNA editing is required to restore their gene function. Our RT-PCR assays indicated that RNA editing occurs in the orchid because a C to U conversion was verified in the initiation codon of *rpl2* transcripts.[30] However, a C to U conversion was not apparent in the *ndhD* transcripts, which further indicates that *ndhD* is a pseudogene. Furthermore, we did not detect RNA editing in the *ndhC*, *ndhD*, *ndhJ*, and *ndhK* transcripts for repairing the internal stop codon,[30] further implying that the eight variously truncated and frameshifted *ndh* genes are probably pseudogenes. Currently, we have been analyzing the transcripts of 76 protein coding genes from *P. aphrodite* subsp. *formosana* to extensively study the pattern of RNA editing in chloroplast.

8.3.5 *IR contraction and expansion*

The border between the inverted repeat region and the single copy region usually shifts among the various plastome species.[11,26,35] Using tobacco as a representative of dicots, a comparison of the exact IR-border positions and their adjacent genes among tobacco, orchid, maize, rice, and wheat is shown in Fig. 8.2. This comparison further demonstrates

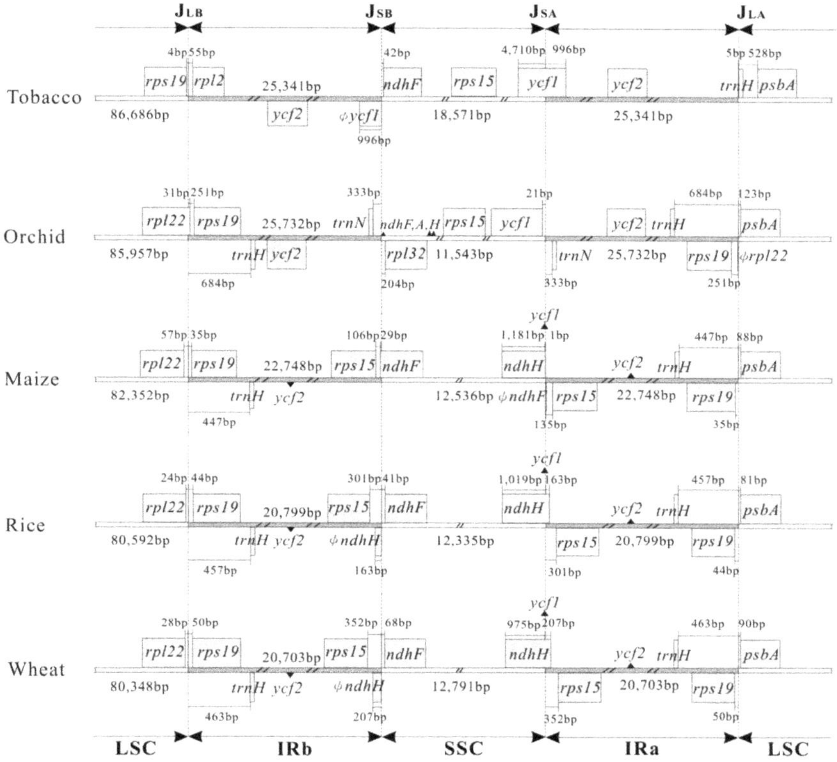

Fig. 8.2. Comparison of border positions of LSC/IR and SSC/IR regions and gene order within two LSC/IR regions among five chloroplast genomes.[30] Various lengths of *ycf1* or *ndhH* pseudogenes are created in tobacco, rice and wheat at the IRb/SSC border. 29 bp of *ndhF* pseudogene is created in maize at the SSC/IRa border. 31 bp of *rpl22* pseudogene is created in *Phalaenopsis aphrodite* subsp. *formosana* at the IRa/LSC border. Both *rps19* and *trnH* genes are located within IR region in *P. aphrodite* subsp. *formosana*, maize, rice, wheat; whereas in tobacco they are located in LSC. The *rps15* gene is located in SSC in *P. aphrodite* subsp. *formosana* and tobacco, whereas it is present in the IR regions in the chloroplast genomes of five grasses. The *ycf1* and *ycf2* genes are lost in five grasses, while *ndhA*, *ndhF*, *ndhH* genes are lost in *P. aphrodite* subsp. *formosana*. Black triangles indicate loss of genes.

that the border positions vary among the chloroplast genomes of angiosperms. As illustrated in Fig. 8.2, the IRb/LSC border in *P. aphrodite* subsp. *formosana* is located within the coding region of *rpl22*, while in maize, rice, and wheat it is located up-stream of the noncoding region of *rpl22*. The IRa/LSC borders in *P. aphrodite* subsp. *formosana*, maize, rice, and wheat are located on the downstream of the non-coding region of *psbA*. The IRa/LSC and IRb/LSC borders of tobacco, however, are located downstream of the noncoding region of *trnH*-GUG and upstream of the non-coding region of *rps19*, respectively. The IRb/LSC borders in many other dicot plants, such as *Panax*,[11] *Spinacia*,[36] *Oenothera*,[37] *Arabidopsis*,[38] *Lotus*,[39] and *Atropa*,[40] are located within the coding region of *rps19*, and as a result, give rise to the *rps19* pseudogenes in IRa. In monocots, the *rps19* and *trnH* genes appear to have been completely shifted into the IR region, which may represent a major structural difference between the chloroplast genomes of monocots and dicots.

In *P. aphrodite* subsp. *formosana*, the IRa/SSC border (Fig. 8.2) is located upstream of the noncoding region of *ycf1*, while in tobacco the border is in the coding region of *ycf1*, with the result that the 996 bp of *ycf1* pseudogene in IRb is created. In grasses, however, the IRa/SSC border is either within the initiation codon of the *ndhH* gene as in maize or shifted into the central region of the *ndhH* gene as in rice and wheat, resulting in the creation of the 163 bp and 207 bp of the *ndhH* pseudogene at the IRb/SSC borders in rice and wheat, respectively. The IRb/SSC border resides downstream of the noncoding region of *rpl32* in *P. aphrodite* subsp. *formosana* and downstream of the coding region of the *ndhF* gene in tobacco, rice and wheat, but in maize it is shifted into the *ndhF* coding region, thereby creating the 29 bp of *ndhF* pseudogene in IRa. The intramolecular recombination of short direct repeat sequences probably is responsible for the shift of the IR/SSC or IR/LSC borders during evolution.[11,26]

The expansions and contractions of the IR region contribute to the variation in length of the chloroplast genome among plant lineages.[11] The IR regions of the grass family (including maize, rice, wild rice, wheat and sugarcane) are shorter than those in *P. aphrodite* subsp. *formosana* and tobacco, mainly because their *ycf2* were lost. The smaller sizes in the SSC regions of the six known monocots are due to loss of the *ycf1* genes in the grass family and the deletion of several *ndh* genes from *P. aphrodite* subsp. *formosana*. In *P. aphrodite* subsp. *formosana* and in dicots, the *rps15* genes are located in the SSC regions, while in the grass family the gene is in the IR regions. Likely, a duplication of the *rps15* gene followed by an inversion occurred in the grass lineage during evolution.

8.4 Conclusion and Prospective

The complete nucleotide sequence of the chloroplast genome of the Taiwan moth orchid (*Phalaenopsis aphrodite* subsp. *formosana*) was determined. The genome contains 76 protein coding genes, four ribosomal RNA genes, 30 tRNA genes, and 24 open reading frames (ORFs). The availability of complete chloroplast genomic sequence of *Phalaenopsis aphrodite* subsp. *formosana* will be very useful in designing molecular markers both for commercial breeding and for phylogenetic classification within Orchidaceae. Furthermore, the plastid genomic information provides the fundamentals to study gene expression and regulation in the chloroplasts of *Phalaenopsis* orchid. In addition, since the lack of complete chloroplast genome sequences is still one of the major limitations for extending chloroplast genetic engineering technology to useful crops, a complete genomic sequence of *Phalaenopsis* chloroplast will provide valuable information. This information can be used for the construction of chloroplast expression vector for site-specific integration of foreign genes through homologous recombination in orchid plastid transformation in the near future.

Acknowledgments

This work was financially supported by grants (NSC92-2317-B006-004) to C.-C. Chang from the National Science Council of Taiwan.

References

1. Martin W, Stoebe B, Goremykin V, *et al.* (1998) Gene transfer to the nucleus and the evolution of chloroplasts. *Nature* **393:**162–165.
2. Palmer JD. (1991) Plastid chromosome: Structure and evolution. In: Vasil IK, Bogorad L (eds.), *Cell Culture and Somatic Cell Genetics in Plants*, Vol. 7A, The Molecular Biology of Plastids. Academic Press, San Diego, pp. 5–53.
3. Martin W, Rujan T, Richly E, *et al.* (2002) Evolutionary analysis of *Arabidopsis*, cyanobacterial, and chloroplast genomes reveals plastid phylogeny and thousands of cyanobacterial genes in the nucleus. *Proc Natl Acad Sci USA* **99:**12246–12251.
4. Ohyama K, Fukuzawa H, Kohchi T, *et al.* (1986) Chloroplast gene organization deduced from complete sequence of liverwort *Marchantia polymorpha* chloroplast DNA. *Nature* **322:**572–574.

5. Shinozaki K, Ohme M, Tanaka M, *et al.* (1986) The complete nucleotide sequence of tobacco chloroplast genome: Its gene organization and expression. *EMBO J* **5:**2043–2049.
6. Raubeson LA, Jansen RK. (2005) Chloroplast genomes of plants. In: Herry RJ (ed.), *Plant Diversity and Evolution: Genotypic and Phenotypic Variation in Higher Plants*. CABI Publishing, Cambridge, pp. 45–68.
7. Sugiura M. (1995) The chloroplast genome. *Essays Biochem* **30:**49–57.
8. Jansen RK, Raubeson LA, *et al.* (2005). Methods for obtaining and analyzing whole chloroplast genome sequences. *Methods Enzymol* **395:**348–384.
9. Asano T, Tsudzuki T, Takahashi S, *et al.* (2004) Complete nucleotide sequence of the sugarcane (*Saccharum officinarum*) chloroplast genome: A comparative analysis of four monocot chloroplast genomes. *DNA Res* **11:**93–99.
10. Calsa T, Carraro DM, Benatti MR, *et al.* (2004) Structural features and transcript-editing analysis of sugarcane (*Saccharum officinarum* L.) chloroplast genome. *Curr Genet* **46:**366–373.
11. Kim KJ, Lee HL. (2004) Complete chloroplast genome sequences from Korean ginseng (*Panax schinseng Nees*) and comparative analysis of sequence evolution among 17 vascular plants. *DNA Res* **11:**247–261.
12. Tang J, Xia H, Cao M, *et al.* (2004) A comparison of rice chloroplast genomes. *Plant Physiology* **135:**412–420.
13. Kugita M, Yamamoto Y, Fujikawa T, *et al.* (2003) RNA editing in hornwort chloroplasts makes more than half the genes functional. *Nucleic Acids Res* **31:**2417–2423.
14. Wolf PG, Rowe CA, Hasebe M. (2004) High levels of RNA editing in a vascular plant chloroplast genome: Analysis of transcripts from the fern *Adiantum capillus-veneris*. *Gene* **339:**89–97.
15. Burrows PA, Sazanov LA, Svab Z, *et al.* (1998) Identification of a functional respiratory complex in chloroplasts through analysis of tobacco mutants containing disrupted plastid *ndh* genes. *EMBO J* **17:**868–876.
16. Drescher A, Ruf S, Calsa T, *et al.* (2000) The two largest chloroplast genome-encoded open reading frames of higher plants are essential genes. *Plant J* **22:**97–104.
17. Daniell H, Khan MS, Allison L. (2002) Milestones in chloroplast genetic engineering: An environmentally friendly era in biotechnology. *Trends Plant Sci* **7:**84–91.
18. Atwood JT. (1986) The size of the Orchidaceae and the systematic distribution of epiphytic orchids. *Selbyana* **9:**171–186.
19. Chang SB, Chen WH, Chen HH, *et al.* (2000) RFLP and inheritance patterns of chloroplast DNA in intergeneric hybrids of *Phalaenopsis* and *Doritis*. *Botany Bulletin of Academia Sinica* **41:**219–223.

20. Cafasso D, Widmer A, *et al.* (2005) Chloroplast DNA inheritance in the orchid *Anacamptis palustris* using single-seed polymerase chain reaction. *J Hered* **96:**66–70.
21. Neyland R, Urbatsch LE. (1996) The *ndhF* chloroplast gene detected in all vascular plant divisions. *Planta* **200:**273–277.
22. Cameron KM. (2004) Utility of plastid *psaB* gene sequences for investigating intrafamilial relationships within Orchidaceae. *Mol Phylogenet Evol* **31:**1157–1180.
23. Robinson SP, Downton WJ. (1984) Potassium, sodium, and chloride content of isolated intact chloroplasts in relation to ionic compartmentation in leaves. *Arch Biochem Biophys* **228:**197–206.
24. Stewart CN Jr, Via LE. (1993) A rapid CTAB DNA isolation technique useful for RAPD fingerprinting and other PCR applications. *Biotechniques* **14:**748–750.
25. Hiratsuka J, Shimada H, Whittier R, *et al.* (1989) The complete sequence of the rice (*Oryza sativa*) chloroplast genome: Intermolecular recombination between distinct tRNA genes accounts for a major plastid DNA inversion during the evolution of the cereals. *Mol Gen Genet* **217:**185–194.
26. Maier RM, Neckermann K, Igloi GL, Kossel H. (1995) Complete sequence of the maize chloroplast genome: Gene content, hotspots of divergence and fine tuning of genetic information by transcript editing. *J Mol Biol* **251:**614–628.
27. Ogihara Y, Isono K, Kojima T, *et al.* (2002) Structural features of a wheat plastome as revealed by complete sequencing of chloroplast DNA. *Mol Genet Gen* **266:**740–746.
28. Shahid Masood M, Nishikawa T, Fukuoka S, *et al.* (2004) The complete nucleotide sequence of wild rice (*Oryza nivara*) chloroplast genome: First genome wide comparative sequence analysis of wild and cultivated rice. *Gene* **340:**133–139.
29. Cafasso D, Pellegrino G, Musacchio A, *et al.* (2001) Characterization of a minisatellite repeat locus in the chloroplast genome of *Orchis palustris* (Orchidaceae). *Curr Genet* **39:**394–398.
30. Chang CC, Lin HC, *et al.* (2006) The chloroplast genome of *Phalaenopsis aphrodite* (Orchidaceae): Comparative analysis of evolutionary rate with that of grasses and its phylogenetic implications. *Mol Biol Evol* **23:**279–291.
31. Wakasugi T, Tsudzuki J, Ito S, *et al.* (1994) Loss of all *ndh* genes as determined by sequencing the entire chloroplast genome of the black pine *Pinus thunbergii*. *Proc Natl Acad Sci USA* **91:**9794–9798.
32. Neyland R, Urbatsch LE. (1996) Phylogeny of subfamily Epidendroideae (Orchidaceae) inferred from *ndhF* chloroplast gene sequences. *Amn J Bot* **83:**1195–1206.

33. Hoch B, Maier RM, Appel K, *et al.* (1991) Editing of a chloroplast mRNA by creation of an initiation codon. *Nature* **353:**178–180.
34. Neckermann K, Zeltz P, Igloi GL, *et al.* (1994) The role of RNA editing in conservation of start codons in chloroplast genomes. *Gene* **146:**177–182.
35. Goulding SE, Olmstead RG, Morden CW, Wolfe KH. (1996) Ebb and flow of the chloroplast inverted repeat. *Mol Gen Genet* **252:**195–206.
36. Schmitz-Linneweber C, Maier RM, Alcaraz JP, *et al.* (2001) The plastid chromosome of spinach (*Spinacia oleracea*): Complete nucleotide sequence and gene organization. *Plant Mol Biol* **45:**307–315.
37. Hupfer H, Swiatek M, Hornung S, *et al.* (2000) Complete nucleotide sequence of the *Oenothera elata* plastid chromosome, representing plastome I of the five distinguishable euoenothera plastomes. *Mol Gen Genet* **263:**581–585.
38. Sato S, Nakamura Y, Kaneko T, *et al.* (1999) Complete structure of the chloroplast genome of *Arabidopsis thaliana*. *DNA Res* **6:**283–290.
39. Kato T, Kaneko T, Sato S, *et al.* (2000) Complete structure of the chloroplast genome of a legume, *Lotus Japonicus*. *DNA Res* **7:**323–330.
40. Schmitz-Linneweber C, Regel R, Du TG, *et al.* (2002) The plastid chromosome of *Atropa belladonna* and its comparison with that of *Nicotiana tabacum*: The role of RNA editing in generating divergence in the process of plant speciation. *Mol Biol Evol* **19:**1602–1612.

Chapter 9

Analysis of Expression of *Phalaenopsis* Floral ESTs

Wen-Chieh Tsai[†], *Yu-Yun Hsiao*[‡], *Zhao-Jun Pan*[‡] *and Hong-Hwa Chen*[*,‡]

Orchids have great diversity of specialized pollination and ecological strategies and provide a rich setting for studying evolutionary relationships and molecular biology. The sophisticated orchid flower morphology offers an opportunity to discover new variant genes and different levels of complexity in the morphogenesis of flowers. To obtain plentiful gene information from orchid reproductive organs, we constructed a cDNA library of mature flower buds of *Phalaenopsis equestris*, a native diploid species of *Phalaenopsis* in Taiwan. A total of 5,593 expressed sequence tags (ESTs) from randomly selected clones were identified and characterized. Cluster analysis enabled the identification of a unigene set of 3,688 sequences. The abundance of transcripts with predicted cellular roles were functionally characterized by using the BLASTX matches to known proteins. Comparison of the relative EST frequencies based on functional categories among floral tissues of five species including *P. equestris*, *Acorus Americanus*, *Asparagus officinalis*, *Oryza sativa*, and *Arabidopsis thaliana* was performed. The most highly transcribed genes in *Phalaenopsis* floral buds are those coding for RNA-dependent RNA polymerase of *Cymbidium* mosaic virus, followed by heat shock protein genes. A total 217 putative transcription factor related ESTs were identified. C3H and trihelix

*Corresponding author.

[†]*Department of Biological Science and Technology, Chung Hwa University of Medical Technology, Tainan County, Taiwan.*

[‡]*Department of Life Sciences, National Cheng Kung University, Tainan, Taiwan.*

families occupied 25% of the transcribed transcription factor genes, indicating that the profile of the transcription factors in orchid flower buds is polarized. The extensive analysis of the genes in floral organs adds to the growing repertoire of known plant genes and may also reveal unique features of the reproductive organs of orchids.

9.1 Introduction

9.1.1 *Arabidopsis and rice structure genomics*

Access to a complete, finished genome for any organism provides the basis for large-scale exploration of biology. With respect to plants, *Arabidopsis* has secured the historical record of being the first plant genome to be sequenced (Arabidopsis Genome Initiative, 2000),[1] with rice (*Oryza sativa*) coming in second.[2,3] The *Arabidopsis* genome is essentially complete with the exception of a few gaps, primarily at the centromeres. The rice genome was completed in December 2004 by a public effort of the International Rice Genome Sequencing Project (IRGSP; http://rgp.dna.affrc.go.jp/IRGSP/). The value of both organisms as model species for plant biology is further supported by the availability of not one genome sequence, but multiple genome sequences. For *Arabidopsis*, the public consortium sequenced to the draft level the heavily utilized Columbia accession, while a private company, Cereon, sequenced the second most utilized accession, Landsberg erecta (Ler).[4] For rice, there have been four genome sequencing efforts, three being focused on the Nipponbare cultivar from the temperate subspecies japonica[2,5,6] and one on the 93–11 variety from the tropical subspecies indica.[3] In addition to the nuclear genomes, both the chloroplast and mitochondrial genomes of *Arabidopsis* and rice are publicly available.[7–9] Both *Arabidopsis* and rice genome sequencing was preceded by expressed sequence tag (EST) sequencing, as this provides not only an inexpensive sampling method for the expressed fraction of a genome, but also a quantitative profile of the expression levels in specific tissues. ESTs also have utilities, as the cDNA clones themselves are valuable reagents for functional genomic studies. Currently, there are 326,202 *Arabidopsis* and 299,998 rice ESTs in the dbEST division of GenBank (http://www.ncbi.nlm.nih.gov/dbEST/dbESTsummary.html).

9.1.2 *The significance of Orchidaceae for plant science*

The family of Orchidaceae has an estimated 17,000 to 35,000 species, making it among the largest families of flowering plants.[10] The species are known for their diversity of both specialized pollination and ecological strategies and provide a rich subject for investigating evolutionary relationships and molecular biology. The versatility and specialization in orchid floral morphology, structure, and physiological properties have fascinated botanists and collectors for centuries. Their most astonishing evolution is seen in their reproductive biology. Like most angiosperm flowers, orchids have two whorls of perianth segments. The outer whorl consists of three sepals. Three petals make up the inner whorl, one of them (the labellum or lip) being highly evolved with different size, form, color, and general appearance from the other two. In some genera, such as *Paphiopedilum* and *Phragmipedium*, the labellum has even evolved into a pouch. In all genera, the male and female reproductive organs are fused to form a gynostemium. Modifications of the perianth, androecium, and gynoecium can represent a basis for a variety of floral morphology. The highly sophisticated flower organization offers the opportunity to discover new variant genes and different levels of complexity in the morphogenetic networks.[11,12]

9.2 Expressed Sequence Tags

9.2.1 *Significance of EST in genome research*

Single-run partial sequencing of randomly selected clones is a widely used tool in genome research. Expressed sequence tags (ESTs) have played significant roles in accelerating gene discovery, including gene family expansion,[13,14] large-scale expression analysis,[15–17] and elucidating phylogenetic relationships.[17] Libraries of cDNAs are routinely prepared and contain tens of thousands of clones, representing a variety of specific tissue types and a snapshot of gene expression during defined developmental stages and following specific biotic and abiotic challenges. Recent developments in high volume biotechnology combined with advanced DNA sequencing technology have made it feasible to perform large-scale EST sequencing projects.

The concept of using cDNAs as a route to expedite gene discovery was first demonstrated in the early 1980s.[19] In 1990, Sydney Brenner proposed that an obvious method for characterizing the "important" part of the human genome would involve looking at messengers from the expressed genes, thus advocating the application of high-throughput methods for transcriptome sampling.[20] Mark Adams first used the term EST in relation to gene discovery and the human genome project in 1991.[21] Currently there are near 25 million ESTs in the National Center for Biotechnology Information (NCBI) public collection-dbEST database (http://www.ncbi.nlm.nih.gov/dbEST/index.html), representing a wide taxonomic variety of fungi, plants, and animals. With many large-scale EST sequencing projects in progress and new projects being initiated, the number of ESTs in the public domain will likely increase substantially, providing additional opportunities for intra- and inter-specific expression comparisons on a genomics scale.

Expressed sequence tags are created by partially sequencing randomly isolated gene transcripts that have been converted into cDNA. EST sequencing initially favored the 5′-end of directionally cloned cDNAs because the 5′-sequences are likely to contain more protein coding sequence than the 3′-ends, which often contain significant untranslated regions (UTRs). Currently, the 3′-end of the cDNA clone is often preferred because it is likely to offer more unique sequence (in many cases, the UTR) and can be used to distinguish between gene paralogues. EST sequencing strategies in which both ends of the cDNA are sequenced are also becoming prevalent. In the absence of complete genome sequences, the cDNA (and its EST) remains the only link back to the genome. However, cDNA sequencing remains one of the more accessible and widely used methods for sampling the actively transcribed portion of the gene space. The preparation of cDNA libraries depends on the underlying mRNA population of a cell, tissue, or organism. The genomic structure of the host plant is, therefore, largely inconsequential. EST data have been directly applied for gene discovery,[22,23] evaluation of the genome-wide gene content and structure,[24] as well as *in silico* comparative expression analysis between different plant tissues.[16,17] Moreover, ESTs can be a valuable resource for highthroughput expression analysis via the cDNA-array technology.[25–28] Recent studies have revealed that mRNA transcripts can also contain repeat motifs, and the abundance of ESTs available makes this an attractive potential source of microsatellite markers.[29] EST-derived SSRs have been developed for many plants, such as *Triticum aestivum* L.,[30] *Hordeum vulgare* L.,[31] *Gossypium* L.,[32] and *Medicago truncatula*.[33]

9.2.2 *Application of full-length cDNAs (FL-cDNA) to functional genomics*

Most of the EST projects are based on cDNA libraries in which most of the inserts are not full-length. ESTs are useful for making a catalogue of expressed genes, but not for further study of gene function. The only viable alternatives to EST sequencing that address the attributes of incomplete sequence coverage and nucleotide quality are the full-length cDNA sequences. Consequently, genome-scale collections of the full-length cDNAs of expressed genes are both essential for gene identification and annotation of complex genomes and provide a valuable resource for experimentalists interested in gene function. In addition, full-length cDNAs are useful resources for transgenic analysis, such as overexpression, antisense suppression, double-stranded RNA interference (dsRNAi), and biochemical analysis to study the function of the encoded proteins.[34] Thus, there has been a large effort to produce annotated EST and full-length cDNA collections for mammals and plants.[35–37]

Full-length cDNA sequences are obtained by Cap-Trapper or other technologies that take advantage of the property of the cap structure, and sequencing clones for both 5′- and 3′-end.[38–44] Such a strategy yields many individual ESTs that can be assembled into a single contig and greatly improves identification of the complete exon structures of eukaryotic genes.[45]

9.2.3 *Bioinformatics of plant EST collections*

It has been shown previously that EST databases are valid and reliable sources of gene expression data.[15–17] With the rapid expansion of available EST data, opportunities for digital gene expression analysis will continue to expand. As a result of advances in computational molecular biology and biostatistics, it is possible to mine and analyze large-scale EST datasets efficiently and exhaustively.[15–17]

Bioinformatics-based sequence resources have been developed that address the quality, redundancy, and partial nature of EST sequences. The first crucial step in adding value to EST sequences involves clustering and assembling the ESTs into a more manageably sized dataset of better quality clustered sequences. Sequences are aggressively trimmed of vector and polylinker remnants before a fast clustering method places the ESTs into buckets of similar sequences. A final assembly step places the clustered sequences into logical contigs and singletons.[46,47] The clustered sequences are typically longer than any individual EST and are of a higher quality.

These cluster consensus and singleton sequences form the core sequence data within several plant specific EST derived databases. Most of these sequence databases have added further value to the sequences by attaching additional annotation to the sequences and by providing methods to select specific sequences or groups of sequences that satisfy specific criteria. The most valuable annotations and methods are those that assign a tentative function and allow retrieval and identification of sequences on the basis of tissue or challenge specificity.

9.2.4 *Reasons for applying ESTs to study Phalaenopsis equestris*

Based on morphological characteristics, the genus *Phalaenopsis* comprises approximately 45 species that are grouped into nine sections.[10] They have broad geographical distribution as well as commercial value as floricultural commodities. Two native species of *Phalaenopsis*, *P. equestris* and *P. aphrodite* var. formosa[48] have been reported in Taiwan. Few molecular studies of *Phalaenopsis* orchids' reproductive biology exist because of their long life cycles and inefficient transformation systems. With the use of flow cytometry, nuclear DNA contents of *P. equestris* were estimated to be 1.6×10^9 bp ($2n = 2x = 38$), about fourfold that of the rice genome.[49,50] The expansion of plant genomes has mainly been the result of multiplications of retrotransposon repeat sequences.[51] Thus, expressed sequence tags (ESTs) avoid the highly repetitive DNA that make up the bulk of most plant genomes and offer both a reasonable cost and rapid generation of data that can be exploited for gene discovery and comparative genomics.

9.3 Characterization and Significance of *Phalaenopsis* Floral EST

9.3.1 *Significance of ESTs shared no significant similarity to any other protein sequences in public databases*

To provide considerable sequence information from orchid, we characterized the 5,593 floral bud ESTs reading from the 5′-end. These ESTs were assembled and generated by the PHRAP program. Among them, 2,637 ESTs were assembled into 732 contigs. These contigs and 2,956 singletons represented up to 3,688 different genes. The results reflected 34.1% (1 – number of unigenes/number of ESTs) redundancy rate of the library. The unigene set was used for similarity searches and annotation using BLAST algorithm. Putative functions of the cDNAs were assigned after applying a stringency of BLASTX scores >50. More than 75% of the database-matched unigenes were found to match proteins from plants (Table 9.1). By contrast, 8.32% of the unigenes represented genes that had been previously found in non-plant organisms, such as *E. coli*, *Drosophila*, fungi, mice, and humans. Furthermore, about 15% of the unigenes revealed no hits, indicating that these sequences may have specific roles in orchids. Although the deposited sequences in public databases are increasing exponentially year by year, recent EST analysis in non-model plants always showed a substantial portion of expressed genes that are species specific.[52–54] Defining the functional identities of these unidentified genes might advance the understanding of the differences among plant species.

Table 9.1. Overview of the Unigenes BlastX Analysis[a]

Source	No. of Unigenes	Percentage of Total ESTs (%)
Plant	2,833	76.82
Non plant	307	8.32
No hits found	548	14.86
Total	3,688	100

[a] Adapted from Tsai *et al.*, *Plant Science* **170** (2006): 426–432.

9.3.2 *Significance of highest expressing level for transcripts of the Phalaenopsis flower buds*

The sequences in the *Phalaenopsis* flower bud dbEST were further characterized by the functional category of the plant genes.[55] Details of the gene species included in each category are given in Fig. 9.1. A significant proportion of ESTs (15%) exhibited similarity to genes that encode enzymes of primary and secondary metabolism. In addition, the *Phalaenopsis* flower bud cDNA library contains transcripts encoding proteins involved in subcellular organization (6%), transcription (5%), signal transduction related genes (4%), protein fate (4%) and cell cycle (3%), tags to transport (3%) and cell rescue (3%), as well as genes involved in proteins with binding function (3%), protein synthesis (3%), interaction with the cellular environment (1%), transposable element

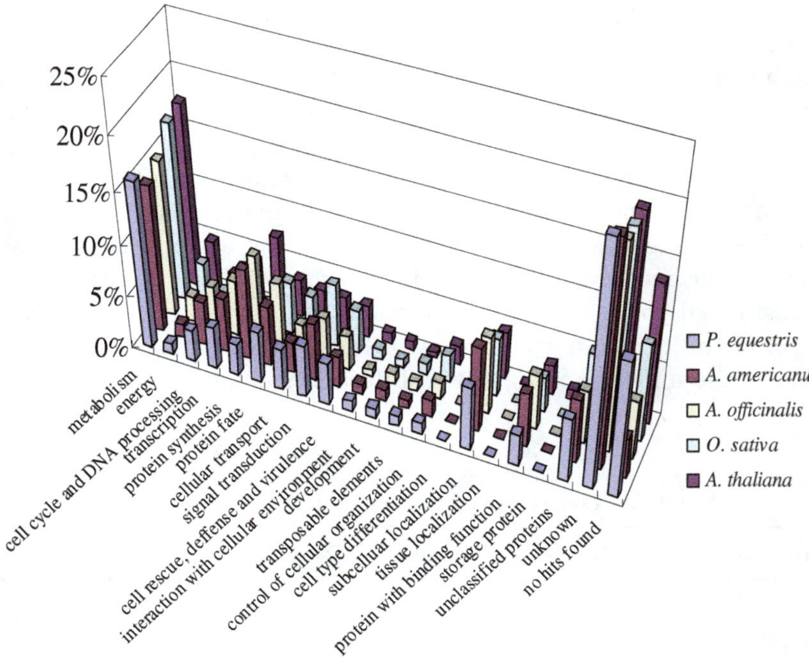

Fig. 9.1. Classification of the BLAST results of the floral ESTs from *P. equestris*, *A. americanus*, *A. officinalis*, *O. sativa* and *A. thaliana*. Relative frequencies of ESTs assigned to predicted functions were similar across these species.

(Adapted from Tsai *et al.*, *Plant Science* **170** (2006): 426–432.)

Table 9.2. Highly Abundant Transcripts Detected in the *P. equestris* Flower Bud dbEST[c]

Putative Function	Organism	No. of ESTs[a]	Percentage[b] (%)
RNA-dependent RNA polymerase	*Cymbidium mosaic virus*	111	1.98
heat shock cognate protein Hsc70-1	*Dictyostelium discoideum*	73	1.40
beta-1,3-glucanase	*Oryza sativa*	48	0.86
40S ribosomal protein S5	*Oryza sativa*	45	0.80
lipid transfer protein	*Arabidopsis thaliana*	44	0.79
phospholipid transfer protein	*Aerides japonica*	36	0.64
metallothionein-like protein	*Musa acuminata*	30	0.54
lipid transfer protein	*Nicotiana glauca*	27	0.48
mannose-binding lectin precursor	*Cymbidium hybrid*	24	0.43
S-adenosyl-L-methionine synthetase	*Dendrobium crumenatum*	20	0.36

[a] Number of clones assigned to the related gene function.
[b] Abundant genes/total number *100.
[c] Adapted from Tsai *et al.*, *Plant Scinece* **170** (2006): 426–432.

(1%), cellular organization (1%), energy production (1%), and development (1%). Another 30% showed similarity to sequences of unknown proteins (24%) and unclassified functions (6%).

These ESTs that have been obtained from a non-normalized cDNA library could be used to reveal global gene expression patterns as deduced from transcript abundance. The highly expressed transcripts of the *Phalaenopsis* flower buds are listed in Table 9.2. The highest expressing level for transcripts of the *Phalaenopsis* flower buds was homologous to RNA-dependent RNA polymerase (1.98%) of *Cymbidium mosaic virus*. Although no virus infection symptoms were observed, these transcripts existed significantly in the floral buds, indicating that experimental materials used here were infected with the virus. Correlated with the highly expressed virus RNA-dependent RNA polymerase transcripts, ESTs related to β-1,3-glucanase were also comparably detected (0.86%). Considerable evidence suggested that β-1,3-glucanase is induced during response of plants to viral, bacterial, and fungal pathogens.[56] In the top 10 highly abundant ESTs, three different kinds of ESTs related to lipid transfer proteins occupied at least 1.91% of the total ESTs. These lipid transfer protein related genes, including lipid transfer protein *per se*, its precursor, and phospholipid transfer proteins play roles in defense and pollen tube adhesion in plants.[57–59] ESTs homologous to heat shock cognate protein Hsc70-1 genes also presented a significant expression level (1.4%). Hsp70s function as molecular chaperones. In addition to heat stress, these proteins can also be induced by other environmental stresses, including cold, drought, or salinity, as well as during various developmental processes, such as embryogenesis, germination, and fruit development.[60–62] Dissection of these gene functions might be a useful direction for further study of orchid floral development.

9.3.3 *Significance of similar expression patterns in floral tissues of flowering plants*

For comparison of floral transcriptome among different plant species, plentiful EST resources deposited at public databases were downloaded to the local server for *in silico* comparative analysis. In particular, ESTs of floral tissue were chosen from *Arabidopsis thaliana*, the model plant of eudicots, *Oryza sativa*, the first monocot whose genome was completely sequenced (International Rice Genome Sequencing Project, 2005);[63] *Acorus americanus*, the basal monocot which provides support as a sister to all other monocots, and *Asparagus officinalis*, an economically important non-grain monocot with the highly derived floral structures found in the grasses and grass relatives.[64] The number of unigene sets were 2,707, 3,934, 4,548, and 4,267 from assembled floral ESTs of *Arabidopsis*, rice, *A. americanus*, and *A. officinalis*, respectively, and functionally characterized and compared to that of *P. equestris*. The results showed that the relative frequencies of ESTs assigned to predicted functions were very similar across these species (Fig. 9.1). Interestingly, transcripts related to cell type differentiation, tissue localization, and storage protein were barely detected in the floral tissues of these five species (Fig. 9.1). These findings suggested that the basic floral transcriptomes were parallel among the flowering plants from basal monocots to monocots and to eudicots.

9.3.4 *Significance of transcription factor expression profiles in the transcriptome of Phalaenopsis flower buds*

Because transcription factors control the expression of a genome and play important roles in all aspects of a higher plant's life cycle, we characterized the transcription factor associated ESTs from the transcriptome of *P. equestris* flower bud through the use of *Arabidopsis* transcription factor sequences as queries.[65] In the orchid flower bud dbEST, 217 ESTs encoding putative transcription factors were identified, occupying 4% (217/5593) of flower bud transcriptome. The most abundantly expressed transcription factor gene families, C3H and trihelix, accounted for a full 25% (54/217) of the overall transcription factor expression (Fig. 9.2). In addition, bHLH (9%), C2H2 (8%) and WRKY (6%) families were found in turn (Fig. 9.2). These five families occupied approximately 50% of the expressed transcription factors. However, BBR-BPC, CCAAT-DR1, CCAAT-HAP2, CPP, GRF, SBP, VOZ-9, and WHIRLY families were not detected. The results suggested that expression of transcription factors in mature orchid flower buds is highly regulated.

The C3H family has been reported to be involved in *Arabidopsis* embryogenesis[66] and response to abscisic acid in *Craterostigma plantagineum*.[67] To our knowledge, documents describing the relationship between C3H

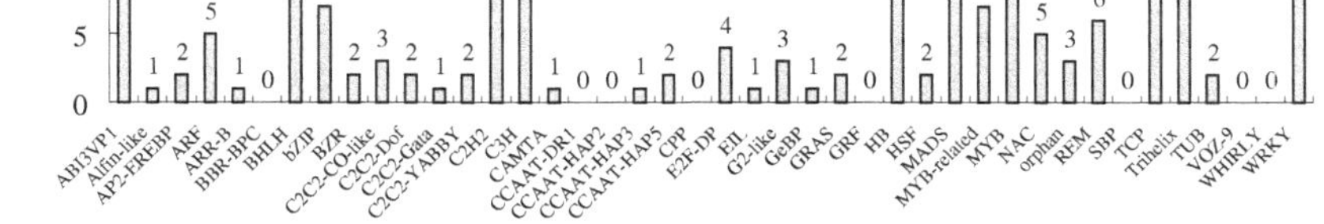

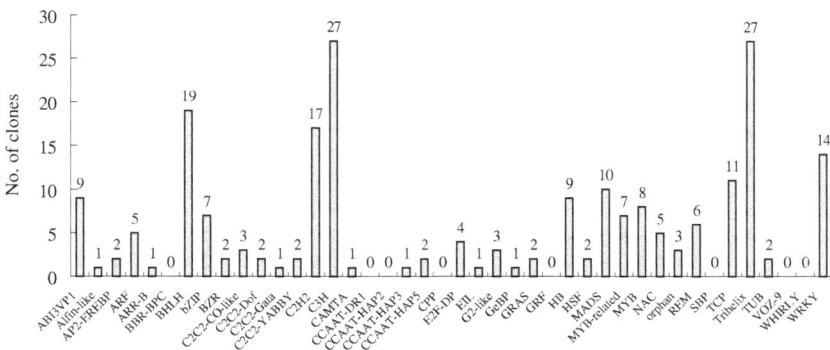

Fig. 9.2. Number of ESTs related to transcription factors appearing in each transcription factor family. A total of 1,583 putative *Arabidopsis* transcription factors were searched against *P. equestris* flower bud dbEST, and then the target ESTs were classified into corresponding transcription factor families. Each bar represents the number of EST clones in one family.

(Adapted from Tsai *et al.*, *Plant Science* **170** (2006): 426–432.)

and floral development have not been proposed. In addition to involvement in light signaling, the trihelix family also regulates perianth architecture in *Arabidopsis* flowers.[68,69] Understanding these gene functions will advance our understanding of orchid floral development. However, there are eight transcription factor families, including BBR-BPC, CCAAT-DR1, CCAAT-HAP2, CPP, GRF, SBP, VOZ-9, and WHIRLY, which were not discovered in our floral bud dbEST. A CPP family gene, *TSO1*, modulates cytokinesis and cell expansion in *Arabidopsis* flowers.[70] The SBP family controls the flowering time, leaf organogenesis, and pollen sac development and is associated with floral development in silver birch.[71–73] These two families of transcription factors might either be rarely expressed genes or might not appear in stage IV flower buds so that EST sampling would not detect them. While no literature concerning the functions of the other six transcription factor families exists, the coincidence of absence of expression of these six families of transcription factors in orchid flower dbEST and the lack of functions in floral development for these transcription factors may explain the precise regulation of transcription factor expression in orchid flowers.

9.4 Conclusions

Large-scale EST sequencing provides a gateway into the genome of organisms owing to the massive information buried in the genome-scale expression data. So far, less than 150 orchid protein sequences have been deposited in public databases (http://www.ncbi.nlm.nih.gov). The random and partial sequencing of orchid cDNA has been a highly rewarding program, particularly because genomic information is still lacking in orchid. The data presented here represents a significant contribution to the publicly accessible expressed sequences for the Orchidaceae family. Application of this knowledge through the common language of nucleotide sequences will provide many candidate genes involved in orchid floral development for further characterization.

References

1. Arabidopsis Genome Initiative. (2000) Analysis of the genome sequence of the flowering plant *Arabidopsis thaliana*. *Nature* **408:**796–815.
2. Goff SA, Ricke D, Lan TH, *et al*. (2002) A draft sequence of the rice genome (*Oryza sativa* L. ssp. *japonica*). *Science* **296:**92–100.

3. Yu J, Hu S, Wang J, *et al.* (2002) A draft sequence of the rice genome (*Oryza sativa* L. ssp. *indica*). *Science* **296:**79–92.
4. Jander G, Norris SR, Rounsley SD, *et al.* (2002) *Arabidopsis* map-based cloning in the post-genome era. *Plant Physiol* **129:**440–450.
5. Sasaki T, Burr B. (2000) International Rice Genome Sequencing Project: The effort to completely sequence the rice genome. *Curr Opin Plant Biol* **3:**138–141.
6. Barry GF. (2001) The use of Monsanto draft rice genome sequence in research. *Plant Physiol* **125:**1164–1165.
7. Unseld M, Marienfeld JR, Brandt P, Brenicke A. (1999) The mitochondria genome of *Arabidopsis thaliana* contains 57 genes in 366,924 nucleolides. *Nat Genet* **15:**59–61.
8. Sato S, Nakamura Y, Kaneko T, *et al.* (1999) Complete structure of the chloroplast genome of *Arabidopsis thaliana*. *DNA Res* **6:**283–290.
9. Notsu Y, Masood S, Nishikawa T, *et al.* (2002) The complete sequence of the rice (*Oryza sativa* L.) mitochondrial genome: Frequent DNA sequence acquisition and loss during the evolution of flowering plants. *Mol Genet Genomics* **268:**434–445.
10. Dressler RL. (1993) *Phylogeny and Classification of the Orchid Family*. Dioscorides Press, Portland, Oregon.
11. Tsai WC, Kuoh CS, Chuang MH, *et al.* (2004) Four *DEF*-like MADS-box genes displayed distinct floral morphogenetic roles in *Phalaenopsis* orchid. *Plant Cell Physiol* **45:**831–844.
12. Tsai WC, Lee PF, Chen HI, *et al.* (2005) *PeMADS6*, a *GLOBOSA/PISTILLATA*-like gene in *Phalaenopsis equestris* involved in petaloid formation, and correlated with flower longevity and ovary development. *Plant Cell Physiol* **46:**1125–1139.
13. Bourdon V, Naef F, Rao PH, *et al.* (2002) Genomic and expression analysis of the 12p11–p12 amplicon using EST arrays identifies two novel amplified and overexpressed genes. *Cancer Res* **62:**6218–6223.
14. Rogaev E, Sherrington R, Rogaeva E, *et al.* (1995) Familial Alzheimers-disease in kindreds with missense mutations in a gene on chromosome-1 related to the Alzheimers-disease type-3 gene. *Nature* **376:**775–778.
15. Ewing RM, Kahla AB, Poirot O, *et al.* (1999) Large-scale statistical analyses of rice ESTs reveal correlated patterns of gene expression. *Genome Res* **9:**950–959.
16. Ogihara Y, Mochida K, Nemoto Y, *et al.* (2003) Correlated clustering and virtual display of gene expression patterns in the wheat life cycle by large-scale statistical analyses of expressed sequence tags. *Plant J* **33:**1001–1011.
17. Ronning C, Stegalkina S, Ascenzi R, *et al.* (2003) Comparative analyses of potato expressed sequence tag libraries. *Plant Physiol* **131:**419–429.

18. Nishiyama T, Fujita T, Shin IT *et al.* (2003) Comparative genomics of Physcomitrella patens gametophytic transcriptome and Arabidopsis thaliana: Implication for land plant evolution. *Proc Natl Acad Sci USA* **100:**8007–8012.
19. Putney SD, Herlihy WC, Schimmel P. (1983) A new troponin T and cDNA clones for 13 different muscle proteins, found by shotgun sequencing. *Nature* **302:**718–721.
20. Brenner S. (1990) The human genome: The nature of the enterprise. *CIBA Found Symp* **149:**6–17.
21. Adams MD, Kelley JM, Gocayne JD, *et al.* (1991) Complementary DNA sequencing: Expressed sequence tags and human genome project. *Science* **252:**1651–1656.
22. Ohlrogge J, Benning C. (2000) Unraveling plant metabolism by EST analysis. *Curr Opin Plant Biol* **3:**224–228.
23. Sterky F, Regan S, Karlsson J, *et al.* (1998) Gene discovery in the wood-forming tissues of poplar: Analysis of 5,692 expressed sequence tags. *Proc Nat Acad Sci USA* **95:**13330–13335.
24. van der Hoeven R, Ronning C, Giovannoni J, *et al.* (2002) Deductions about the number, organization, and evolution of genes in the tomato genome based on analysis of a large expressed sequence tag collection and selective genomic sequencing. *Plant Cell* **14:**1441–1456.
25. Potokina E, Sreenivasulu N, Altschmied L, *et al.* (2002) Differential gene expression during seed germination in barley (*Hordeum vulgare* L.). *Funct Integr Genomics* **2:**28–39.
26. Schena M, Shalon D, Davis RW, Brown PO. (1995) Quantitative monitoring of gene expression patterns with a complementary DNA microarray. *Science* **270:**467–470.
27. Sreenivasulu N, Altschmied L, Panitz R, *et al.* (2002) Identification of genes specifically expressed in maternal and filial tissues of barley caryopses: A cDNA array analysis. *Mol Genet Genomics* **266:**758–767.
28. Sreenivasulu N, Altschmied L, Radchuk V, *et al.* (2004) Transcript profiles and deduced changes of metabolic pathways in maternal and filial tissues of developing barley grains. *Plant J* **37:**539–553.
29. Kantety RV, Rota ML, Matthews DE, Sorrells ME. (2002) Data mining for simple sequence repeats in expressed sequence tags from barely, maize, rice, sorghum and wheat. *Plant Mol Biol* **48:**501–510.
30. Gao LF, Jing RL, Huo NX, *et al.* (2004) One hundred and one new microsatellite loci derived from ESTs (EST-SSRs) in bread wheat. *Theor Appl Genet* **108:**1392–1400.
31. Thiel T, Michalek W, Varshney RK, Graner A. (2003) Exploiting EST databases for the development and characterization of gene-derived SSR-markers in barley (*Hordeum vulgare* L.). *Theor Appl Genet* **106:** 411–422.

32. Saha S, Karaca M, Jenkins JN, *et al.* (2003) Simple sequence repeats as useful resources to study transcribed genes of cotton. *Euphytica* **130:** 355–364.
33. Eujayl I, Sledge MK, Wang L, *et al.* (2004) *Medicago truncatula* EST-SSRs reveal cross-species genetic markers for *Medicago* spp. *Theor Appl Genet* **108:**414–422.
34. Seki M, Satou M, Sakurai T, *et al.* (2004) RIKEN *Arabidopsis* full-length (RAFL) cDNA and its applications for expression profiling under abiotic stress conditions. *J Exp Bot* **55:**213–223.
35. Okazaki Y, Furuno M, Kasukawa T, *et al.* (2002) Analysis of the mouse transcriptome based on functional annotation of 60,770 full length cDNAs. *Nature* **420:**563–573.
36. Nagata T, Iizumi S, Satoh K, *et al.* (2004) Comparative analysis of plant and animal calcium signal transduction element using plant full-length cDNA data. *Mol Biol Evol* **21:**1855–1870.
37. Ota T, Suzuki Y, Nishikawa T, *et al.* (2004) Complete sequencing and characterization of 21,243 full-length human cDNAs. *Nat Genet* **36:**40–45.
38. Carninci P, Kvam C, Kitamura A, *et al.* (1996) High-efficiency full-length cDNA cloning by biotinylated CAP trapper. *Genomics* **37:**327–336.
39. Carninci P, Westover A, Nishiyama Y, *et al.* (1997) High efficiency selection of full-length cDNA by improved biotinylated cap trapper. *DNA Res* **4:**61–66.
40. Carninci P, Nishiyama Y, Westover A, *et al.* (1998) Thermostabilization and thermoactivation of thermolabile enzymes by trehalose and its application for the synthesis of full length cDNA. *Proc Natl Acad Sci USA* **95:**520–524.
41. Suzuki Y, Yoshitomo-Nakagawa K, Maruyama K, *et al.* (1997) Construction and characterization of a full length-enriched and a 5′-end-enriched cDNA library. *Gene* **200:**149–156.
42. Seki M, Carninci P, Nishiyama Y, *et al.* (1998) High-efficiency cloning of Arabidopsis full-length cDNA by biotinylated CAP trapper. *Plant J* **15:**707–720.
43. Carninci P, Hayashizaki Y. (1999) High-efficiency full-length cDNA cloning. *Methods Enzymol* **303:**19–44.
44. Mizuno Y, Carninci P, Okazaki Y, *et al.* (1999) Increased specificity of reverse transcription priming by trehalose and oligo-blockers allows high-efficiency window separation of mRNA display. *Nucleic Acids Res* **27:**1345–1349.
45. Castelli V, Aury JM, Jaillon O, *et al.* (2004) Whole genome sequence comparisons and "full-length" cDNA sequences: A combined approach to evaluate and improve *Arabidopsis* genome annotation. *Genome Res* **14:**406–413.
46. Gordon D, Abajian C, Green P. (1998) Consed: A graphical tool for sequence finishing. *Genome Res* **8:**195–202.

47. Huang X, Madan A. (1999) CAP3: A DNA sequence assemblyprogram. *Genome Res* **9:**868–877.
48. Chen WH, Wang YT. (1996) *Phalaenopsis* orchid culture. *Taiwan Sugar* **43:**11–16.
49. Lin S, Lee HC, Chen WH, *et al.* (2001) Nuclear DNA contents of *Phalaenopsis* species and *Doritis pulcherrima. J Amer Soc Horti Sci* **126:**195–199.
50. Kao YY, Chang SB, Lin TY, *et al.* (2001) Differential accumulation of heterochromatin as a cause for karyotype variation in *Phalaenopsis* orchids. *Ann Bot* **87:**387–395.
51. Rudd S. (2003) Expressed sequence tags: Alternative or complement to whole genome sequences? *Trends Plant Sci* **8:**321–329.
52. Aziz N, Paiva NL, May GD, Dixon RA. (2005) Transcriptome analysis of alfalfa glandular trichomes. *Planta* **221:**28–38.
53. Chen L, Zhao LP, Gao QK. (2005) Generation and analysis of expressed tags from the tender shoots cDNA library of tea plant (*Camellia sinensis*). *Plant Sci* **168:**359–363.
54. Sterky F, Bhalerao RR, Unneberg P, *et al.* (2004) A *Populus* EST resource for plant functional genomics. *Proc Natl Acad Sci USA* **101:**13951–13956.
55. Frishman D, Albermann K, Hani J, *et al.* (2001) Functional and structural genomics using PEDANT. *Bioinformatics* **17:**44–57.
56. Bucher GL, Tarina C, Heinlein M, *et al.* (2001) Local expression of enzymatically active calss I β-1, 3-glucanase enhances symptoms of TMV infection in tobacco. *Plant J* **28:**361–369.
57. Buhot N, Douliez JP, Jacquemard A, *et al.* (2001) A lipid transfer protein binds to a receptor involved in the control of plant defense responses. *FEBS Lett* **509:**27–30.
58. Park SY, Jauh GY, Mollet JC, *et al.* (2000) A lipid transfer-like protein is necessary for lily pollen tube adhesion on *in vitro* stylar matrix. *Plant Cell* **12:**151–163.
59. Wang ST, Wu JH, Ng TB, *et al.* (2004b) A non-specific lipid transfer protein with antifungal and antibacterial activities from the mung bean. *Peptides* **25:**1235–1242.
60. Almoguera C, Prieto-Dapena P, Jordano J. (1998) Dual regulation of a heat shock promoter during embryogenesis: Stage-dependent role of heat shock elements. *Plant J* **13:**437–446.
61. Sun W, van Montagu M, Verbruggen N. (2002) Small heat shock proteins and stress tolerance in plants. *Biochim Biophys Acta* **19:**1–9.
62. Wehmeyer N, Hernandez LD, Finkelstein RR, Vierling E. (1996) Synthesis of small heat-shock proteins is part of the developmental program of late seed maturation. *Plant Physiol* **112:**747–757.
63 International Rice Genome Sequencing Project. (2005) The map-based sequence of the rice genome. *Nature* **436:**793–800.

64 Albert VA, Soltis DE, Carlson JE, *et al.* (2005) Floral gene resources from basal angiosperms for comparative genomics research. *BMC Plant Biol* **5:**5.
65. Riechmann JL, Heard J, Martin G, *et al.* (2002) *Arabidopsis* transcription factors: Genome-wide comparative analysis among eukaryotes. *Science* **290:**2105–2110.
66. Li Z, Thomas TL. (1998) *PEI1*, an embryo-specific zinc finger protein gene required for heart-stage embryo formation in *Arabidopsis*. *Plant Cell* **10:**383–398.
67. Hilbricht T, Salamini F, Bartels D. (2002) CpR18, a novel SAP-domain plant transcription factor, binds to a promoter region necessary for ABA mediated expression of the *CDeT27-45 gene* from the resurrection plant *Craterostigma plantagineum* Hochst. *Plant J* **31:**293–303.
68. Brewer PB, Howles PA, Dorian K, *et al.* (2004) *PETAL LOSS*, a trihelix transcription factor gene, regulates perianth architecture in the *Arabidopsis* flower. *Development* **131:**4035–4045.
69. Wang R, Hong G, Han B. (2004a) Transcript abundance of *rml1*, encoding a putative GT1-like factor in rice, is up-regulated by *Magnaporthe grisea* and down-regulated by light. *Gene* **324:**105–115.
70. Hauser BA, He JQ, Park SO, Gasser CS. (2000) TSO1 is a novel protein that modulates cytokinesis and cell expansion in *Arabidopsis*. *Development* **127:**2219–2226.
71. Cardon G, Hohmann S, Klein J, *et al.* (1999) Molecular characterization of the *Arabidopsis* SBP-box genes. *Gene* **237:**91–104.
72. Lannenpaa M, Janonen I, Holtta-Vuori M, *et al.* (2004) A new SBP-box gene *BpSPL1* in silver birch (*Betula pendula*). *Physiol Plant* **120:**491–500.
73. Unte US, Sorensen AM, Pesaresi P, *et al.* (2003) *SPL8*, an SBP-box gene that affects pollen sac development in *Arabidopsis*. *Plant Cell* **15:**1009–1019.

Chapter 10

Orchid MADS-Box Genes Controlling Floral Morphogenesis

Wen-Chieh Tsai†*, Chin-Wei Lin*‡*, Chang-Sheng Kuoh*‡ *and Hong-Hwa Chen**,‡

Orchids are known for both their floral diversity and ecological strategies. The versatility and specialization in orchid floral morphology, structure, and physiological properties have fascinated botanists for centuries. In floral studies, MADS-box genes contributing to the now-famous "ABCDE model" of floral organ identity control have dominated conceptual thinking. The sophisticated orchid floral organization offers an opportunity to discover new variant genes and levels of complexity different from the ABCDE model. Recently, several remarkable researches involving orchid MADS-box genes have revealed the important roles of these genes in orchid floral development. Knowledge about MADS-box genes encoding ABCDE functions in orchids will provide insights into the highly evolved floral morphogenetic networks of orchids.

10.1 Introduction

With more than 270,000 known species, the angiosperms are by far the most diverse and widespread group of plants. The ancestry of angiosperms is still uncertain. The fossil records showed that they appeared at the

*Corresponding author.

†*Department of Biological Science and Technology, Chung Hwa University of Medical Technology, Tainan County, Taiwan.*

‡*Department of Life Sciences, National Cheng Kung University, Tainan, Taiwan.*

early Cretaceous period, about 130 million years ago. By the end of the Cretaceous period, 65 million years ago, the angiosperms had radiated and become the dominant plants on earth, as they are today. The origin and diversification of angiosperms, what Charles Darwin characterized as "an abominable mystery," has been the subject of much speculation over the last 100 years.[1–3] The rapid explosion in diversity that followed their origin in the early Cretaceous period may be linked to modularity within a new structure, the flower.[4] The flower is the defining reproductive adaptation of angiosperms and is the predominant source of characters for angiosperm taxonomy and phylogeny reconstruction.[5]

Over the past decades, the codification of rigorous methods of phylogenetic analysis, the emergence of molecular techniques, and a renewed interest in developmental pathways during the growth of plant organs have improved the understanding of the relations among angiosperms.[6–9] The angiosperms consist of some small relic basal clades (basal angiosperms), magnollids, and two main clades, monocots and eudicots (Fig. 10.1). The basal angiosperms and magnollids share some primitive traits, such as a typical spiral rather than whorled

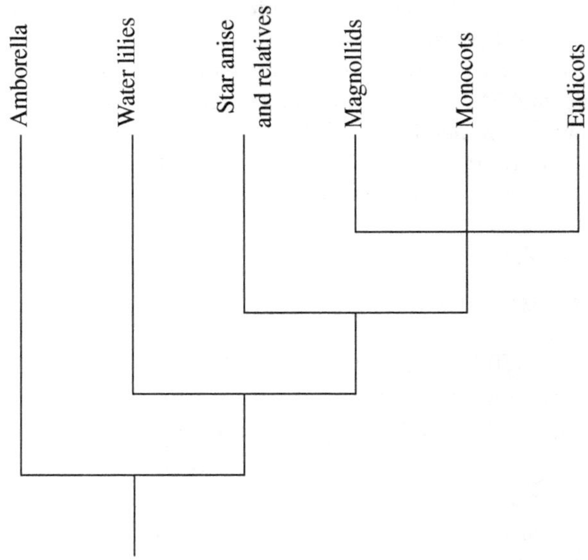

Fig. 10.1. Phylogeny of angiosperms.

arrangement of flower organs.[10,11] The monocots show extreme variation in floral form, including bilaterally symmetric (zygomorphic) flowers with elaborately modified perianth parts. The organization of flower parts is slightly less variable in the core eudicots.

The MADS-box-containing transcriptional regulators have been the focus of floral organ specification, development, and evolutionary studies in plants.[12–14] In the well-known "ABC model," the organ identity in each whorl is determined by a unique combination of three activities by floral organ identity genes, A, B, and C.[12] In any one of the four flower whorls, the expression of A alone specifies sepal formation. The combination of AB determines the development of petals, and the combination of BC offers stamen formation. The expression of the C function alone determines the development of carpels. The functions of A and C are mutually repressive.[15]

The ABC genes were cloned from a wide range of species, and the model has been used to explain floral organ development in plants.[15–21] From studies of *Petunia*, the ABC model was later extended to include D-class genes, which specify ovules.[22] An important recent discovery was that another set of MADS-box genes, *SEPALLATA1*, *2*, and *3*, function redundantly to specify petals, stamens, and carpels as well as floral determinacy.[23] Recently, *SEP4* has been defined. *sep1 sep2 sep3 sep4* quadruple mutants develop vegetative leaves rather than sepals, petals, stamens, or carpels.[24] SEPALLATA function or E function has led to a revision of the ABC model into the "ABCDE model."[25,26] The diversification of MADS-box genes during evolution has been proposed to be a major driving force for floral diversity in land plant architecture.[15,27]

Containing more than 20,000 species, the *Orchidaceae*, of class Liliopsida and order Asparagales, is one of the largest angiosperm families. Associated with this enormous size is extraordinary floral diversity. Orchids are extremely rich in species, and speciation rates are presumed to be exceptionally high.[28] Although this spectacular diversification is often thought to be linked to the intimate and sometimes bizarre interaction of many species with their pollinators,[29] we still face the challenge of explaining how these mechanisms work and why they have evolved.

According to the classic view, the orchid flower is composed of five whorls of three segments, each including two perianth whorls, two staminal whorls, and one carpel whorl (Fig. 10.2A).[30] This formation conforms to the general flower structure of many other monocotyledonous

Fig. 10.2. Flowers from wild type (**A**) and peloric mutant (**B**) of *Phalaenopsis equestris*. Bar = 1 cm.

families. *Orchidaceae* represents an unusually coherent group among monocots, possessing several reliable floral morphological synapomorphies, including the presence of a gynostemium, or column, fused by the style and at least part of the androecium, a highly evolved petal called the labellum, and resupination caused by 180° torsion of the pedicel.[31] Within the monocots, only well-known crop species, such as rice and maize have been studied thoroughly, but the highly reduced flowers make these plants unsuitable for general floral development studies. All expected whorls in the flowers are present in orchids, and the highly sophisticated flower organization offers an opportunity to discover new variant genes and different levels of complexity within morphogenetic networks. Therefore, the *Orchidaceae* can be used to validate the ABC model in the monocots and study how MADS-box genes are involved in defining the different, highly specialized structures in orchid flowers.

10.2 *Phalaenopsis* Orchid Floral Ontogeny

Floral ontogenetic studies provide an essential tool for research into systematics and phylogenetics as well as developmental regulation. Kurzweil and Kocyan[32] reviewed research on species that have been studied for orchid floral ontogeny and included data on the total number of genera and species in Dressler's subfamilies of *Orchidaceae* and the number of species whose floral ontogeny was studied, for a total of 128 genera and 211 species in five subfamilies examined. The authors concluded that the sequence of floral organ initiation on the floral apex and the timing are identical in all monandrous species studied, including *Phalaenopsis*.[32]

Our recent studies of floral ontogeny in *Phalaenopsis equestris*, a native species in Taiwan, could strengthen the evidence for monandrous orchid floral ontogeny (Fig. 10.3). The floral primordium of this species emerges on the flanks of the inflorescence meristem and then enlarges (Fig. 10.3A and 3B). The first organs initiated are the lateral sepals located at the adaxial side (Fig. 10.3C). The primordium of the lip soon follows (Fig. 10.3D). Then the lateral petal primodia emerge (Fig. 10.3E). The last observed tepal is the dorsal sepal located at the abaxial side (Fig. 10.3F). Subsequently, the first organ of the gynostemium

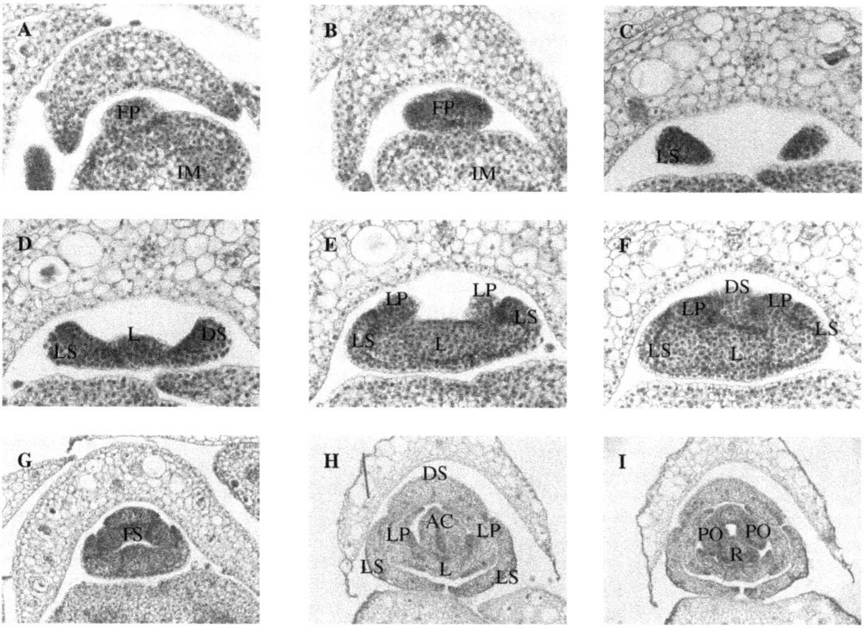

Fig. 10.3. Transverse sections of developing *P. equestris* flowers showing floral ontogeny. (**A**) and (**B**) Initiation of floral primodium. (**C**)–(**F**) Differentiation of floral perianth. (**G**) Initiation of gynostemium. (**H**) and (**I**) Late developmental stage of gynostemium. FP: floral primodium, IM: inflorescence meristem, LS: lateral sepal, L: lip, LP: lateral petal, DS: dorsal sepal, FS: fertile stamen, AC: anther cap, PO: pollinium, R: rostellum.

to be initiated is the fertile stamen (Fig. 10.3G). Then, the carpel apices become visible in the central part of the receptacle. The anther cap and the rostellum derived from the median stigma lobe develop later (Fig. 10.3H and 3I). This process of floral development in *P. equestris* is nearly parallel to that of the other monandrous orchids. The detailed knowledge gained from orchid floral ontogeny will provide a solid foundation for the study of morphogenetic functions of orchid floral organ identity genes.

10.3 A-Class Genes in Orchids

To date, almost all cloned orchid MADS-box genes involved in floral development were from Epidendroideae, the largest orchid subfamily with many more genera and species than all of the other four subfamilies together (Table 10.1). So far, four A-class genes were identified from *Dendrobium*. One of them, *DOMADS2*, was isolated from *Dendrobium* grex Madame Thong-In.[33] The other three genes were cloned from *Dendrobium thyrsiflorum* and named *DthyrFL1*, *DthyrFL2*, and *DthyrFL3*.[34] The *APETALA1/FRUITFULL* (*AP1/FUL*) MADS-box genes lineage within the eudicots was recognized as two clades (the *euAP1* and *euFUL* clades), while the non-core eudicots and monocots have sequences similar only to *euFUL* genes.[35] Sequence analysis showed that these genes contain the C-terminal *FUL*-like motif LPPWML of monocot FUL-like proteins, but this motif is not present in the sequence of *DthyrFL3*.[34] Phylogenetic analysis showed that *DthyrFL1* and *DOMADS2* are orthologous genes (Fig. 10.4A). The existence of *DthyrFL2* and *DthyrFL3* represents a recent duplication event in *D. thyrsiflorum* (Fig. 10.4A).[34] *DOMADS2* is expressed in both the shoot apical meristem and the emerging floral primordium throughout the process of floral transition and later in the column of mature flowers.[33] The expression pattern of *DOMADS2*, from the shoot apical meristem and increasing in later stages of floral development, suggests that *DOMADS2* is one of the earliest regulatory genes during the transition of flowering. *DthyrFL* genes are expressed during inflorescence development and also in developing ovules.[34] These A-class genes in orchid may be involved in floral meristem identity and column and ovule development. Unlike its homolog *AP1* in *Arabidopsis*, *DOMADS2* may not be associated with the development of the first

Table 10.1. Floral MADS-box Genes and their Expression in Orchidaceae

Class	Gene	Species	Expression Location									Analysis
			Sepal	Petal	Labellum	Column	Pedicel/ Ovary	Root	Leaf	Shoot	Ovule	
A	*DOMADS2*	*Dendrobium* grex Madame Thong-In	–[a]	–	–	+[b]	–	–	–	+	ND[c]	I[d], II[e]
	DthyrFL1	*Dendrobium thyrsiflorum*	ND	ND	ND	ND	ND	–	–	ND	+	III[f]
	DthyrFL2	*Dendrobium thyrsiflorum*	ND	ND	ND	ND	ND	–	–	ND	+	III
	DthyrFL3	*Dendrobium thyrsiflorum*	ND	ND	ND	ND	ND	–	–	ND	+	III
B (paleo*AP3*)	*OMADS3*	*Oncidium* Gower Ramsey	+	+	+	+	ND	ND	+	ND	ND	IV[g]
	PeMADS2	*Phalaenopsis equestris*	+	+	–	–	–	–	–	–	ND	I
	PeMADS3	*Phalaenopsis equestris*	–	+	+	–	–	–	–	–	ND	I
	PeMADS4	*Phalaenopsis equestris*	–	–	+	+	–	–	–	–	ND	I
	PeMADS5	*Phalaenopsis equestris*	+	+	+	–	–	–	–	–	ND	I
	DcOAP3A	*Dendrobium crumenatum*	+	+	+	+	+	ND	+	ND	ND	II, IV
	DcOAP3B	*Dendrobium crumenatum*	–	+	+	+	+	ND	–	ND	ND	IV

(*Continued*)

Table 10.1. (*Continued*)

Class	Gene	Species	Expression Location									Analysis
			Sepal	Petal	Labellum	Column	Pedicel/ Ovary	Root	Leaf	Shoot	Ovule	
B (PI)	*PeMADS6*	*Phalaenopsis equestris*	+	+	+	+	+	–	–	–	ND	I, II
	DcOPI	*Dendrobium crumenatum*	+	+	+	+	+	ND	–	ND	ND	II, IV
	ORCPI	*Orchis italica*	ND	ND	ND	ND	ND	ND	ND	ND	ND	
C	*PeMADS1*	*Phalaenopsis equestris*	–	–	–	+	+	–	–	–	ND	I, IV
	PhalAG1	*Phalaenopsis Hatsuyuki*	–	–	–	+	+	ND	ND	ND	+	II, IV
	DthyrAG1	*Dendrobium thyrsiflorum*	–	–	–	+	ND	ND	ND	ND	+	II, III
	DcOAG1	*Dendrobium crumenatum*	+	+	+	+	+	ND	–	ND	ND	II, IV
D	*PhalAG2*	*Phalaenopsis Hatsuyuki*	–	–	–	+	+	ND	ND	ND	+	II, IV
	DthyrAG2	*Dendrobium thyrsiflorum*	–	–	–	+	ND	ND	ND	ND	+	II, III
	DcOAG2	*Dendrobium crumenatum*	–	–	–	+	+	ND	–	ND	ND	II, IV

(*Continued*)

Table 10.1. ***(Continued)***

			Expression Location									
Class	**Gene**	**Species**	**Sepal**	**Petal**	**Labellum**	**Column**	**Pedicel/ Ovary**	**Root**	**Leaf**	**Shoot**	**Ovule**	**Analysis**
E	*OM1*	*Aranda* Deborah	+	+	+	–	ND	ND	ND	ND	ND	I
	DOMADS1	*Dendrobium* grex Madame Thong-In	+	+	+	+	+	–	–	–	ND	I, II
	DOMADS3	*Dendrobium* grex Madame Thong-In	–	–	–	–	+	–	–	–	–	I, II
	DcOSEP1	*Dendrobium crumenatum*	+	+	+	+	+	ND	–	ND	ND	IV

[a]Transcripts of gene not detected.
[b]Transcripts of gene detected.
[c]Detection of transcripts has not been performed.
[d]Transcripts detected by northern blotting.
[e]Transcripts detected by *in situ* hybridization.
[f]Transcripts are detected by real-time RT-PCR.
[g]Transcripts detected by RT-PCR.

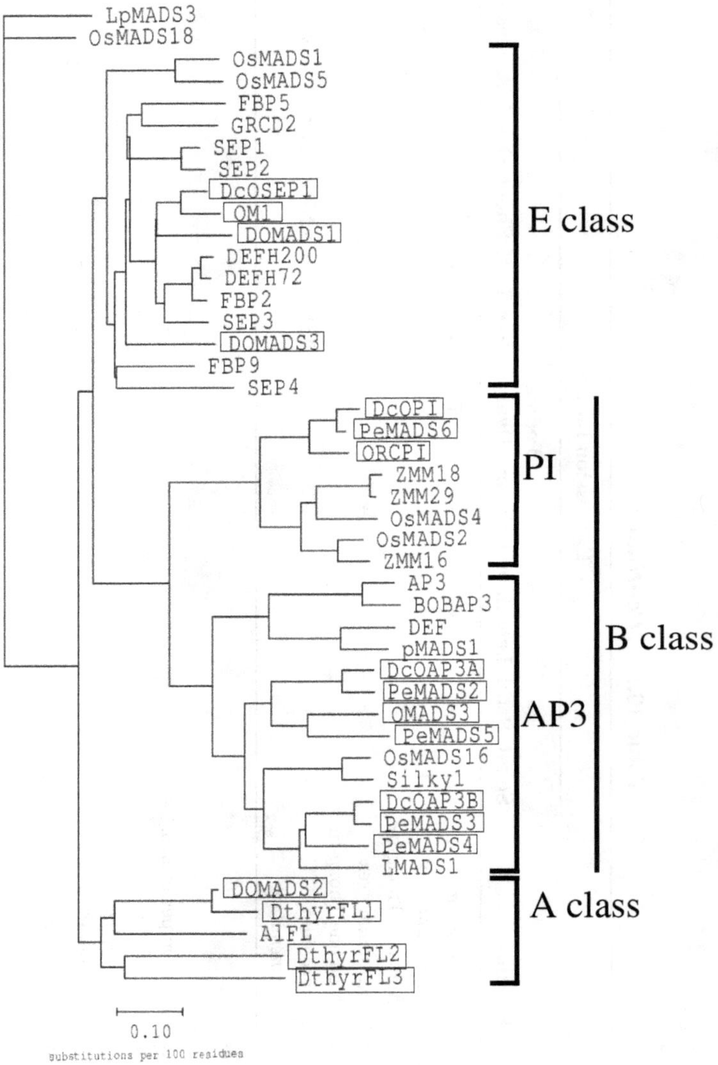

(A)

Fig. 10.4. Phylogenetic relationship of MADS-box genes of ABCDE class. (**A**) Phylogenetic analysis of A-, B- and E-class genes. (**B**) Phylogenetic analysis of C- and D-class genes. Orchid MADS-box genes are highlighted by open boxes.

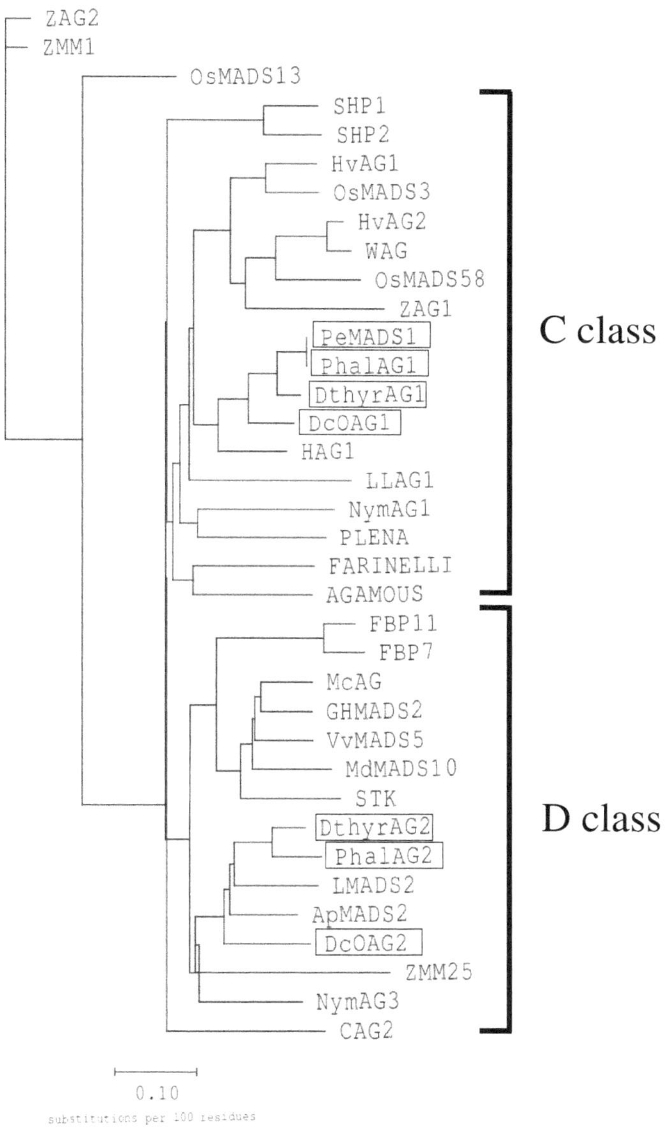

(B)

Fig. 10.4. (*Continued*)

two whorls. However, whether the *DthyrFL* genes associate with perianth formation is not clear. In addition, a MADS-box gene *OMADS1* in *Oncidium* is also involved in floral initiation and formation and belongs to the *AGL6* subfamily rather than the A-class genes.[36] Transgenic *Arabidopsis* and tobacco overexpressing *OMADS1* showed significantly reduced plant size, flowering extremely early, and losing inflorescence indeterminacy.[36]

10.4 B-Class Genes in Orchids

Both the developmental and the biochemical aspects of B-class genes required to specify the identity of petals in whorl 2 and stamens in whorl 3 appear to be conserved in many core eudicots.[15,16] The B-class genes in the monocots rice and maize are similar in function to the core eudicots B-class genes.[19,20,37,38] The orchid flowers have a petaloid perianth arrangement that could be explained by a "modified ABC model" in that the expression of the B-class genes has expanded to whorl 1.[39] In addition, the orchid flowers display an elaborated labellum that is a highly modified petal. Because the function of A-class genes is poorly defined in angiosperms, studies of petal development and evolution have generally focused on B-class genes.[40] The extraordinary floral diversity in orchids may be associated with the evolution of B-class genes.

The molecular functions of B-class genes have been studied in more detail than that of any other floral homeotic genes in orchids. So far, various numbers of *APETALA3* (*AP3*)-like and *PISTILLATA* (*PI*)-like genes have been isolated from several orchids. These include 1 *AP3*-like *OMADS3* from *Oncidium* Gower Ramsey; 4 *AP3*- and 1 *PI*-like genes from *P. equestris*; and 2 *AP3*- and 1 *PI*-like genes from *Dendrobium crumenatum*.[41–43] All these *AP3*-like genes are members of the paleo*AP3* lineage (Fig. 10.4A).

The paleo*AP3* genes identified from orchids were subdivided into two subclades. One subclade contains *OMADS3*, *PeMADS5*, *DcOAP3A,* and *PeMADS2* and the other *OcOAP3B*, *PeMADS3*, and *PeMADS4* (Fig. 10.4A). This division suggests that the ancestor of *Orchidaceae* might have two paleo*AP3*-like genes, and further gene duplication has at least taken place in the *AP3* clade in the monocots. Interestingly, both *OMADS3* and *PeMADS5* do not show a paleoAP3 motif, which suggests that they are orthologous genes. Although they share similar expression patterns in orchid floral organs, *PeMADS5* is not expressed in vegetative tissues, but *OMADS3* can be detected in leaves.[41,42] Phylogenetic analysis also showed that *DcOAP3A* and *PeMADS2* are orthologous genes (Fig. 10.4A). However, they possess different expression patterns. Similar to *OMADS3*, *DcOAP3A* is ubiquitously expressed in all floral organs and leaves, while *PeMADS2* is predominantly expressed in sepals and petals.[41–43] Recently, we discovered at least three paleo*AP3* genes displaying distinct expression patterns in *Oncidium* floral organs (our unpublished data). In addition, we also noticed that the expression profile of *OMADS3* examined by Hsu and Yang[41] was indeed composed of the expression patterns from 2 *paleo*AP3 genes of *Oncidium* (our unpublished data). A specialized paleo*AP3* gene, *PeMADS4*, discovered in *P. equestris*, is specifically expressed in the labellum and column, which suggests that its function is associated with the development of these organs. Gene duplication is important for generating new genes during evolution[44] and, therefore, may lead to the generation of new organs. In orchids, duplication of paleo*AP3*-like genes, followed by diversification and specialization of *PeMADS4*-like genes probably is concomitant with the appearance of a new floral organ, the labellum.

Overexpression of *paleo*AP3 genes from *Oncidium*, *Dendrobium*, and *Phalaenopsis* under the control of the cauliflower mosaic virus 35S promoter has been examined in *Arabidopsis*[41,43] (our unpublished data). Consistently, these results showed that the flower morphology of transgenic *Arabidopsis* plants overexpressing the orchid *paleo*AP3 genes is indistinguishable from that of wild-type plants. Dominant-negative mutation strategy was further used to investigate the functions of *OMADS3* and *DcOAP3A*.[43] By doing this, the *OMADS3* was shown to have a function similar to that of the A functional gene in regulating flower formation as well as floral initiation, while *DcOAP3A* had a putative B function.[41,43] However, these results could not reflect the real roles the genes may play during orchid floral development.

Peloric flowers that are actinomorphic mutants with lip-like petals are widely found in natural populations of species from Veronicaceae, Gesneriaceae, Labiatae, and Orchidaceae.[45] With the high frequency of orchid peloric mutants derived from micropropagation (Fig. 10.2B), we were able to infer the individual roles the diversified *paleo*AP3 genes play in orchids by comparing the expression patterns of the 4 *paleoAP3* genes (*PeMADS2*, *PeMADS3*, *PeMADS4* and *PeMADS5*) in wild-type *Phalaenopsis* floral organs and peloric mutants.[42] First, both *PeMADS2* and *PeMADS5* were expressed in sepals of wild-type plants, but only *PeMADS2* transcript was detected in the sepals of peloric mutants whose morphology is not affected. This result suggests that *PeMADS5* is dispensable, while *PeMADS2* is crucial for sepal development. Second, the expression of all four *PeMADS* genes, except *PeMADS4*, was detected in wild-type petals. However, the expression of *PeMADS5* was not noted in the lip-like petals of peloric mutant. This result suggests that *PeMADS5* is associated with petal development. Third, the expression of *PeMADS4* was concentrated in the lips and columns in wild-type plants and is extended to the lip-like petals in the peloric mutant. Because *PeMADS4* was detected as well in the lip-like petal of the peloric mutant, it might be required for labellum identity. Fourth, *PeMADS3* showed similar expression patterns in the wild-type plant and the peloric mutant, which implies its important function in inner perianth whorl morphogenesis.

So far, only one *PI*-like gene has been found in *D. crumenatum* and *P. equestris*: *DcOPI* and *PeMADS6*, respectively.[43,46] Southern blot hybridization results supported that the *Phalaenopsis* orchid genome contains only one copy of the *PI*-like gene.[46] Both genes are expressed in all the floral organs, except that *PeMADS6* is not detected in the pollinia of *P. equestris*.[43,46] In addition, *PeMADS6* is expressed in the undeveloped ovary.[46] Tsai *et al.*[46] suggested that the expression of *PeMADS6* in the ovary has an inhibitory effect on the development of the ovary, and auxin acts as the candidate signal to regulate the repression of *PeMADS6* expression in the ovary. Furthermore, *PeMADS6* is not differentially expressed in wild-type and peloric floral organs, which suggests that *PeMADS6* is not responsible for the altered phenotype of the peloric mutant. Overexpression of *DcOPI* or *PeMADS6* in *Arabidopsis* demonstrated that both genes share the angiosperm *PI* function.[43,46] Further evidence came from the complementation of the *pi-1* phenotype in *Arabidopsis* by overexpressing *DcOPI* and showed that *DcOPI* is able to substitute *PI* in *Arabidopsis*,[46] while *PeMADS6* could not complement the *pi-4* mutant (our unpublished data).

In conclusion, the expression patterns of B-class genes in orchid floral organs nicely fits the "modified ABC model" in that their expression has expanded to whorl 1 in plants possessing nearly identical morphology of sepals and petals.[39] paleo*AP3* genes are highly duplicated in the Epidendroideae genome. Diversification and fixation of both these gene sequences and expression profiles might cause subfunctionalization and even neofunctionalization. The driving force of the specialized labellum and diversified orchid flowers may be linked to the fast evolution rate of paleo*AP3* genes. Studies of the B-class genes from other orchid subfamilies, such as Apostasioideae, Cypripedioideae, Spiranthoideae, Orchidoideae, and even more members of Epidendroideae,[47] will provide profound knowledge to resolve the mystery of orchid floral development and diversification.

10.5 C- and D-Class Genes in Orchids

A gynostemium or column, comprising stamen filaments adnate to a syncarpous style, is normally regarded as a structure peculiar to orchids.[47] The development of the column, which involves whorls 3 and 4, would be one of the most interesting subjects for elucidating the evolution of C-class genes. In most orchid flowers, ovary and ovule development is precisely and completely triggered by pollination and, therefore, orchids offer a unique opportunity to study D-class genes involved in ovule development. Recently, one C- and one D-class gene were isolated independently from three orchid species: *Phalaenopsis* Hatsuyuki (*PhalAG1*, *PhalAG2*),[48] *D. thyrsiflorum* (*DthyrAG1*, *DthyrAG2*),[49] and *D. crumenatum* (*DcOAG1*, *DcOAG2*).[43] *PhalAG1*, *DthyrAG1*, *DcOAG1* were classified in the C lineage of AG-like genes, with *PhalAG2*, *DthyrAG2*, *DcOAG2* classified in the D lineage (Fig. 10.4B). *PhalAG1*, *PhalAG2*, *DthyrAG1*, and *DthyrAG2* share similar spatial expression patterns in the column, ovary and developing ovules, despite the fact that these four genes belong to different lineages.[48,49] One possible explanation is that C- and D-class genes in orchid would act redundantly with each other in floral and ovule development. Although *PhalAG1* is expressed in all the floral organs at their initiation, its expression quickly decreases and is detected only in the column and ovary of mature flowers, which is also true for *DthyrAG1*.[48,49] We also identified a C-class gene, *PeMADS1*, from *P. equestris*, whose expression patterns were consistent with those of *PhalAG1* and *DthyrAG1* (our unpublished data). However, *DcOAG1* is expressed in all mature floral organs, and *DcOAG2* is expressed in only the anther cap and column of *D. crumenatum*.[43] The unusual expression patterns of *DcOAG1* in monocots evoke that its regulatory mechanism is independently evolved in *D. crumenatum* as in some basal angiosperms, such as *Illicium* and *Persea*,[50] but the function of *DcOAG1* and *Arabidopsis AG* is conserved, as seen by the phenotypic similarity between transgenic *Arabidopsis* expressing either *35S::DcOAG1* or *35S::AG*.[43] The molecular mechanism of morphogenesis of orchid gynostemium is still enigmatic, but mutation of C-class genes in orchids could possibly provide the opportunity to shed light on the mystery.

10.6 E-Class Genes in Orchids

E-class genes are required for floral organ identity in all four orchid floral organs as well as for floral determinacy.[24,51] They have been shown to form ternary complexes with A- and B-class proteins, by a yeast three-hybrid system, and can mediate the interactions between B- and C-class proteins in higher-order complexes.[52] The first E-class gene found, *OM1*, was isolated from the supposed bigeneric hybrid *Aranda* Deborah.[53] The other three E-class genes, *DOMADS1*, *DOMADS3*, *DcOSEP1*, were identified from *Dendrobium*. Two were cloned from *D.* grex Madame Thong-In and the third from *D. crumenatum*.[33,43] Phylogenetic analysis showed that *OM1* was clustered with *DOMADS1* and *DcOSEP1*, with *DOMADS3* separated at a distance from the other three E-class genes (Fig. 10.4A). *DOMADS1* RNA is uniformly expressed in both the inflorescence meristem and floral primordium and later exists in all of the floral organs.[33] The expression pattern of *DOMADS1* in mature flowers coincides with that of its counterpart *DcOSEP1* in *D. crumenatum* and their orthologs in *Arabidopsis*.[23,33,43] However, *OM1* is expressed in mature flowers and not in young developing inflorescence or young floral buds. In mature flowers, it is expressed only in petals and weakly in sepals, but not in the column.[53] Spatiotemporal expression differences imply that functional diversification among these genes closely relates to phylogeny. The onset of *DOMADS3* transcription occurs in the early shoot apical meristem at the stage before the differentiation of the first flower primordium and later can only be detected in the pedicels.[33] *DOMADS3* may function as a regulatory factor not only in early floral transition, but also in the development of the pedicel.

The fact that the expression of E-class genes overlaps with that of ABC genes in orchids suggests that the higher-order MADS complexes are involved in orchid floral development. Recently, one line of evidence that MADS proteins form higher-order complexes comes from the formation of DcOAP3A-DcOPI-DcOSEP1 and DcOAP3B-DcOPI-DcOSEP1 detected by yeast three-hybrid experiments.[43]

10.7 Perspective

Owing to the large genome size, the long life cycle and the inefficient transformation system of orchids, few studies of orchid biology exist. Recently, the genetic architecture of the sophisticated floral organization of orchids has begun to be investigated. Given that orchids represent one of the most successful and diverse plant families worldwide, the development of genomic resources is imperative.[54] Thanks to advanced progress in genomics and bioinformatics, plentiful gene information and integrated bioinformatics tools are ready to be used for studying orchid biology.[55–57] The effort of many scientists will promise to lead to a better understanding of the molecular and genetic mechanisms of orchid floral control in the years to come.

Acknowledgments

We thank Wen-Huei Chen (Department of Life Sciences, National Science Council, National University of Kaohsiung, Kaohsiung, Taiwan) for leading us into the field of orchids; Chang-Sheng Kuoh (Department of Life Sciences, National Cheng Kung University, Tainan, Taiwan) for sharing his knowledge of orchid floral ontogeny; and Tung-Hua David Ho (Institute of Plant and Microbial Biology, Academia Sinica, Taipei, Taiwan) and Michel Delseny (Laboratory of Plant Genome and Development, University of Perpignan, France) for helpful discussions. Grant funding from both the National Science Council and Council of Agriculture in Taiwan supports the current research in our labs.

References

1. Cronquist A. (1988) *The Evolution and Classification of Flowering Plants*, 2nd ed. New York Botanical Garden, Bronx, NY.
2. Darwin C. (1903) *The Different Forms of Flowers on Plants of the Same Species*. D. Appleton, New York.

3. Doyle JA. (1994a) Origin of the angiosperm flower: A phylogenetic perspective. *Plant Sys Evol Supple* **8:**7–29.
4. Carrol SB. (2001) Chance and necessity: The evolution of morphological complexity and diversity. *Nature* **409:**1102–1109.
5. Doyle JJ. (1994b) Evolution of a plant homeotic multigene family: Towards connecting molecular systematics and molecular developmental genetics. *Syst Biol* **43:**307–328.
6. Crane PR, Friis EM, Pedersen KR. (1995) The origin and early diversification of angiosperm. *Nature* **353:**31–37.
7. Soltis PS, Solti DE, Savolainen V, *et al.* (2002) Rate heterogeneity among lineages of tracheophytes: Integration of molecular and fossil data and evidence for molecular living fossils. *Proc Natl Acad Sci USA* **99:**4430–4435.
8. Chaw SM, Chang CC, Chen HL, Li WH. (2004) Dating the monocot-dicot divergence and the origin of core eudicots using whole chloroplast genomes. *J Mol Evol* **58:**424–441.
9. Davies TJ, Barraclough TG, Chase MW, *et al.* (2004) Darwin's abominable mystery: Insights from a supertree of the angiosperms. *Proc Natl Acad Sci USA* **101:**1904–1909.
10. Albert VA, Gustafsson MHG, DiLaurenzio L. (1998) Ontogenetic systematics, molecular developmental genetics, and the angiosperm petal. In: Soltis DE, Soltis PS, Doyle JJ (eds.), *Molecular Systematics of Plants II: DNA Sequencing*. Kluwer Academic Publishers, Boston, pp. 349–374.
11. Endress PK. (2001) Origins of flower morphology. *J Exp Zool* **291:**105–115.
12. Weigel D, Meyerowitz EM. (1994) The ABCs of floral homeotic genes. *Cell* **78:**203–209.
13. Purugganan MD, Rounsley SD, Schmidt RJ, Yanofsky MF. (1995) Molecular evolution of flower development: Diversification of the plant MADS-box regulatory gene family. *Genetics* **140:**345–356.
14. Münster T, Pahnke J, Di Rosa A, *et al.* (1997) Floral homeotic genes were recruited from homologous MADS-box genes preexisting in the common ancestor of ferns and seed plants. *Proc Natl Acad Sci USA* **94:**2415–2420.
15. Theissen G, Becker A, Rosa AD, *et al.* (2000) A short history of MADS-box genes in plants. *Plant Mol Biol* **42:**115–149.
16. Irish VF, Kramer EM. (1998) Genetic and molecular analysis of angiosperm flower development. *Adv Bot Res* **28:**197–230.
17. Kang HG, Jeon JS, Lee S, An G. (1998) Identification of class B and class C floral organ identity genes from rice plants. *Plant Mol Biol* **38:**1021–1029.
18. Kramer EM, Dorit RL, Irish VF. (1998) Molecular evolution of genes controlling petal and stamen development: Duplication and divergence within the *APETALA3* and *PISTILLATA* MADS-box gene lineage. *Genetics* **149:**765–783.

19. Ambrose BA, Lerner DR, Ciceri P, *et al.* (2000) Molecular and genetic analyses of the *silky1* gene reveal conservation in floral organ specification between eudicots and monocots. *Mol Cell* **5:**569–579.
20. Whipple CJ, Ciceri P, Padilla CM, *et al.* (2004) Conservation of B-class floral homeotic gene function between maize and *Arabidopsis*. *Development* **131:**6083–6091.
21. Yamaguchi T, Lee DY, Miyao A, *et al.* (2006) Functional diversification of the two C-class MADS-box genes *OSMADS3* and *OSMADS58* in *Oryza sativa*. *Plant Cell* **18:**15–28.
22. Angenent GC, Colombo L. (1996) Molecular control of ovule development. *Trends Plant Sci* **1:**228–232.
23. Pelaz S, Ditta GS, Baumann E, *et al.* (2000) B and C floral organ identity functions require SEPALLATA MADS-box genes. *Nature* **405:**200–203.
24. Ditta G, Pinyopich A, Robles P, *et al.* (2004) The *SEP4* gene of *Arabidopsis thaliana* functions in floral organ and meristem identity. *Curr Biol* **14:**1935–1940.
25. Theissen G, Saedler H. (2001) Floral quartets. *Nature* **409:**469–471.
26. Zahn LM, Leebens-Mack J, DePamphilis CW, *et al.* (2005) To B or Not to B a flower: The role of *DEFICIENS* and *GLOBOSA* orthologs in the evolution of the angiosperms. *J Hered* **96:**225–240.
27. Irish VF, Litt A. (2005) Flower development and evolution: Gene duplication, diversification and redeployment. *Curr Opin Genet Dev* **15:**454–460.
28. Gill DE. (1989) Fruiting failure, pollinator inefficiency, and speciation in orchids. In: Otte D, Endler JA (eds.), *Speciation and its Consequences*. Sunderland, Sinauer Associates, pp. 458–481.
29. Darwin C. (1885) *On the Various Contrivances by which Orchids are Fertilized by Insects*, 2nd edn. John Murray, London.
30. Brown R. (1810) *Prodromus Florae Novae Hollandiae et Insula Van-Diemen*. J. Johnson and Co., London.
31. Rudall PJ, Bateman RM. (2002) Roles of synorganisation, zygomorphy and heterotopy in floral evolution: The gynostemium and labellum of orchids and other lilioid monocots. *Biol Rev* **77:**403–441.
32. Kurzweil H, Kocyan A. (2002) Ontogeny of orchid flowers. In: Kull T, Arditti J (eds.), *Orchid Biology: Reviews and Perspectives*, VIII. Kluwer, Dordrecht, pp. 83–138.
33. Yu H, Goh CJ. (2000) Identification and characterization of three orchid MADS-box genes of the AP1/AGL9 subfamily during floral transition. *Plant Physiol* **123:**1325–1336.
34. Skipper M, Pedersen KB, Johansen LB, *et al.* (2005) Identification and quantification of expression levels of three *FRUITFULL*-like MADS-box genes from the orchid *Dendrobium thyrsiflorum* (Reichb. f.). *Plant Sci* **169:**579–586.

35. Litt A, Irish VF. (2003) Duplication and diversification in the *APETALA1/FRUITFULL* floral homeotic gene lineage: Implications for the evolution of floral development. *Genetics* **165:**821–833.
36. Hsu HF, Huang CH, Chou LT, Yang CH. (2003) Ectopic expression of an orchid (*Oncidium* Gower Ramsey) *AGL6*-like gene promotes flowering by activating flowering time genes in *Arabidopsis thaliana*. *Plant Cell Physiol* **44:**783–794.
37. Lee S, Jeon JS, An K, *et al.* (2003) Alteration of floral organ identity in rice through ectopic expression of *OsMADS16*. *Planta* **217:**904–911.
38. Nagasawa N, Miyoshi M, Sano Y, *et al.* (2003) *SUPERWOMAN1* and *DROOPING LEAF* genes control floral organ identity in rice. *Development* **130:**705–718.
39. van Tunen AJ, Eikelboom W, Angenent GC. (1993) Floral organogenesis in *Tulipa*. *Flowering Newsl* **16:**33–37.
40. Baum DA, Whitlock BA. (1999) Plant development: Genetic clues to petal evolution. *Curr Biol* **9:**R525–R527.
41. Hsu HF, Yang CH. (2002) An orchid (*Oncidium* Gower Ramsey) *AP3*-like MADS gene regulates floral formation and initiation. *Plant Cell Physiol* **43:**1198–1209.
42. Tsai WC, Kuoh CS, Chuang MH, *et al.* (2004) Four *DEF*-like MADS box genes displayed distinct floral morphogenetic roles in *Phalaenopsis* orchid. *Plant Cell Physiol* **45:**831–844.
43. Xu Y, Teo LL, Zhou J, *et al.* (2006) Floral organ identity genes in the orchid *Dendrobium crumenatum*. *Plant J* **46:**54–68.
44. Ohno S. (1970) *Evolution by Gene Duplication*. Springer, Berlin Heidelberg, New York.
45. Cubas P. (2004) Floral zygomorphy, the recurring evolution of a successful trait. *BioEssays* **26:**1175–1184.
46. Tsai WC, Lee PF, Chen HI, *et al.* (2005) *PeMADS6*, a *GLOBOSA/PISTILLATA*-like gene in *Phalaenopsis equestris* involved in petaloid formation, and correlated with flower longevity and ovary development. *Plant Cell Physiol* **46:**1125–1139.
47. Dressler RL. (1993) *Phylogeny and Classification of the Orchid Family*. Cambridge University Press, Cambridge.
48. Song I-J, Nakamura T, Fukuda T, *et al.* (2006) Spatiotemporal expression of duplicate *AGAMOUS* orthologues during floral development in *Phalaenopsis*. *Dev Genes Evol* **4:**1–13.
49. Skipper M, Johansen LB, Pederson KB, *et al.* (2006) Cloning and transcription analysis of an *AGAMOUS*- and *SEEDSTICK* ortholog in the orchid *Dendrobium thyrsiflorum* (Reichb. f.). *Gene* **366:**266–274.
50. Kim S, Koh J, Yoo MJ, *et al.* (2005) Expression of floral MADS-box genes in basal angiosperms: Implications for the evolution of floral regulators. *Plant J* **43:**724–744.

51. Kaufmann K, Melzer R, Theissen G. (2005) MIKC-type MADS-domain proteins: Structure modularity, protein interactions and network evolution in land plants. *Gene* **347:**183–198.
52. Honma T, Goto K. (2001) Complexes of MADS-box proteins are sufficient to convert leaves into floral organs. *Nature* **409:** 525–529.
53. Lu ZX, Wu M, Loh CS, *et al.* (1993) Nucleotide sequence of a flower-specific MADS box cDNA clone from orchid. *Plant Mol Biol* **23:**901–904.
54. Cozzolino S, Widmer A. (2005) Orchid diversity: An evolutionary consequence of deception? *Trends Ecol Evol* **20:**487–494.
55. Wang HC, Kuo HC, Chen HH, *et al.* (2005) KSPF: Using gene sequence patterns and data mining for biological knowledge management. *Expert Syst Appl* **28:**537–545.
56. Hsiao YY, Tsai WC, Kuoh CS, *et al.* (2006) Comparison of transcripts in *Phalaenopsis bellina* and *Phalaenopsis equestris* (Orchidaceae) flowers to deduce monoterpene biosynthesis pathway. *BMC Plant Biol* **6:**14.
57. Tsai WC, Hsiao YY, Lee SH, *et al.* (2006) Expression analysis of the ESTs derived from the flower buds of *Phalaenopsis equestris*. *Plant Sci* **170:**426–432.

Chapter 11

Pseudobulb-Specific Gene Expression of *Oncidium* Orchid at the Stage of Inflorescence Initiation

Jun Tan[†], *Heng-Long Wang*[‡] *and Kai-Wun Yeh*[*,§]

Oncidium pseudobulb is a critical organ in the developmental stage. It is a sink for nutrition, water, and mineral storage organ during the vegetative growth stage. The growth and development determine Oncidium plant growth cycle transforming from the vegetative into the reproductive stage. Therefore, the genes activating in the pseudobulb of before-inflorescence stage attract much research interest. In this chapter, we present a subtractive EST data bank of EST genes, which are actively expressing in the pseudobulb tissues before inflorescence initiation. In total, 74.8% of 636 unique gene parts were annotated on the database of the NCBI GenBank. Largely, all EST was classified into carbohydrate metabolism involved in mannan, pectin, and starch bosynthesis, transportation, stress-related, and regulatory function. Most of the genes that were differentially expressed are involved in early flowering development, carbohydrate metabolism, and stress-response physiology. The efficient pseudobulb-specific EST-library represented an explicit transcriptome profile before the flowering stage. It is a valuable data bank for molecular biology study in orchid.

*Corresponding author.

[†]*College of Bioinformation, Chongqing University of Post and Telecom, Chongqing, China.*

[‡]*Department of Life Science, National Kaohsiung University, Kaohsiung, Taiwan.*

[§]*Institute of Plant Biology, College of Life Science, National Taiwan University, Taipei, Taiwan.*

11.1 Introduction

Orchids (*Orchidaceae*, L.) are the largest family of plants and are predominantly found in the tropics and subtropics, especially in the mountains of the tropical Americas and Southeast Asia. The number of orchid species may exceed 30,000 and constitute almost 30% of monocots or 10% of flowering plants (Royal Botanic Gardens, Kew, 2003). The genus *Oncidium* has become more and more commercially important in the market for cut flowers and houseplants.[1] Consequently, efforts to improve the economic traits of this ornamental plant are immense at present. After the method of efficient plant regeneration through somatic embryogenesis from *Oncidium* orchid callus,[2] a routine procedure of transformation with *Agrobacterium tumefaciens* was also established.[1] Sweet pepper ferredoxin-like protein (*pflp*) was used as a novel selection marker for orchid transformation.[3] At the same time the selection marker was demonstrated to confer resistance against soft rot disease in *Oncidium* orchid.[4] Other results showed that the function of its AP3-like MADS gene in regulating floral formation and initiation.[5] Recently, a sucrose-phosphate synthase gene highly expressed in *Oncidium* flowers and leaves was cloned.[6]

At the base of the second upper leaf, *Oncidium* orchid has an enlarged bulb-like stem, termed a pseudobulb, which is important for the storage and maintenance of moisture, mineral nutrition, and carbohydrates during both auxiliary bud and flower development.[7] It is clear that current and former pseudobulbs are connected,[8] and the carbohydrate pool in the current pseudobulb varied strikingly during inflorescence development.[9] From initiation to the end of inflorescence development, galactonic acid 1,4-lactone, mannan, and hexoses including glucose, fructose, and galactose gradually decreased in the current pseudobulb, but sucrose and mannose contents almost always remained low. During the period before flowering, there was dramatic accumulation followed by degradation of starch.[10]

Here, subtractive ESTs were generated by subtracting *Rsa* I-digested cDNAs of the upper leaf tissue from those of the pseudobulb. The highly expressed and tissue-specific genes, such as peroxidase, sodium/dicarboxylate cotransporter, mannose binding lectin, senescence, or resistance associated proteins and other genes were thought of as having specific functions in the pseudobulb. Redundancies of these ESTs also revealed some interesting information on their expressed gene family members. The sequence dataset was not on a large scale, but did cover most of the genes involved in the metabolisms of *Oncidium* pseudobulb, mainly carbohydrates, transportation, stress-related, cell cycle, and regulatory function. All of them were related to the water, nutrition, and energy support of the pseudobulb during inflorescence development. Our research on pseudobulb- specific gene expression is beneficial for the further functional determination of these genes and to identify the physiological linkage with floral time and qualities.

11.2 Overall Distribution of Subtractive ESTs from Pseudobulb of Oncidium Gower Ramsy

In total, 1248 clones were sequenced. 1088 clones were readable and had an average length of 1031 bp. The distribution of their length is shown in Fig. 11.1. After the pGEM-T easy vector sequence was removed by cross match from them, 1080 inserts with a length more than 100 bp were accepted as subtractive ESTs for further study. Their length was 430 bp on average and in the range of 100 to 1200 bp, mainly from 200 to 800 bp (Fig. 11.1A).

After phrap, 1080 subtractive ESTs were assembled into 149 contigs and 543 singletons. Furthermore, they were aligned into 51 clusters and 585 singles by blastn. There were 69 contigs in the 585 singles, so 516 real singles had only one subtractive EST, 48.3% of the total ESTs. The number of ESTs in clusters, also called cluster size, reflected the abundance of mRNAs. Including singles with one EST, the distribution of cluster size was indicated in Fig. 11.1B, and clusters were divided

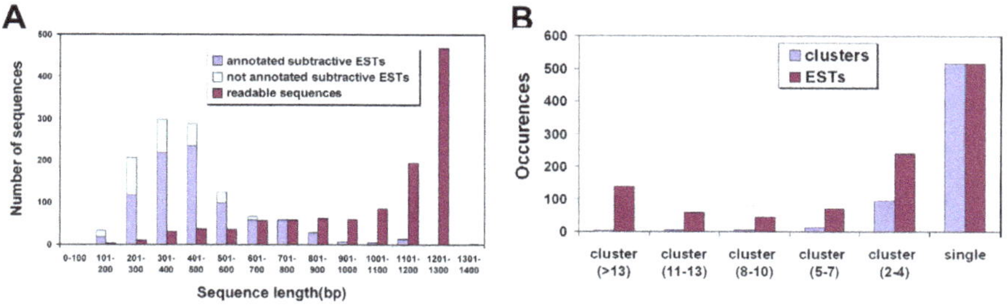

Fig. 11.1. Alignment and annotation of subtractive ESTs. (**A**) Length distribution of readable sequences and subtractive ESTs. (**B**) Prevalence distribution of subtractive EST cluster size.

(From Tan *et al.*, *Biotech Letts* (2005) **27:** 1517–1528. The authors wish to thank *Biotechnology Letters* for providing the figures and tables cited in the text.)

into six classes. There were 95 clusters containing 2–4 ESTs, constituting 22.3% of the total accepted ESTs (241 of 1080 subtractive ESTs) and 79.1% of the total clusters (95 of 120 clusters).

There were 12 clusters containing 5–7 ESTs, constituting 6.5% of the total accepted ESTs (70 of 1080 subtractive ESTs) and 10.0% of the total clusters (12 of 120 clusters). These clusters represented lipid transfer protein, DP-E2F-related protein, senescence-associated protein, mannose-binding lectin, sucrose synthase, peroxidase, GDSL-motif lipase/hydrolase, pollen allergen, and two unknown protein genes. One cluster had no significant similarity to any protein sequence in the GenBank nr database.

There were five clusters containing 8–10 ESTs, constituting 4.1% of the total accepted ESTs (44 of 1080 subtractive ESTs) and 4.1% of the total clusters (5 of 120 clusters). These clusters represented mannose-binding lectin, BURP domain protein, peroxidase, chitinase, and 19 genes. One cluster had no significant similarity to any protein sequence in the GenBank nr database. There were another five clusters containing 11–13 ESTs constituting 5.5% of the total accepted ESTs, 59 of 1080 subtractive ESTs. These clusters represent short-chain dehydrogenase/reductase and two unkonwn genes (*Arabidopsis* T23E23.17, T18N14.110). Two clusters had no significant similarity to any protein sequence in the GenBank nr database.

Three clusters were regarded as the most abundant transcripts. They constituted only 2.5% of the total clusters (3 of 120 clusters), but 12.7% of the total subtractive ESTs (137 of 1080 subtractive ESTs). Two of them were peroxidase genes, and one was a sodium/dicarboxylate cotransporter gene. More details and the possible functions of these genes are shown and discussed in Table 11.1. Comparatively, mannose binding lectin genes were expressed predominately in normal pseudobulb (data not shown).

11.3 Annotation and Classification of the Subtractive ESTs

A total of 636 clusters and singles including 1080 subtractive ESTs were submitted to blastX for homologous searching with the nr database in GenBank. 439 clusters and singles composed of 808 ESTs (484 contigs and singletons) were identified. A total of 309 of them comprising 614 ESTs

Table 11.1. Assembled Clusters that Contain more than 4 ESTs

Gene Annotation	Reference Organism	GI Number	E Value	ESTs
peroxidase (EC 1.11.1.7)	*Gossypium irsutum*	7433087	2.10E-47	92
sodium/dicarboxylate cotransporter	*Arabidopsis thaliana*	15238130	5.10E-43	24
peroxidase (EC 1.11.1.7) 2, cationic	*Glycine max*	7433098	6.10E-26	21
T23E23.17	*Arabidopsis thaliana*	9369404	7.00E-11	12
similar to Arabidopsis thaliana T18N14.110	*Oryza sativa*	13486662	7.10E-29	12
short-chain dehydrogenase/ reductase	*Arabidopsis thaliana*	15224306	1.10E-76	11
No hits found				11
mannose-binding lectin	*Cymbidium hybrid*	2144226	2.10E-51	10
BURP domain protein	*Vigna unguiculata*	7106540	6.00E-12	9
No hits found				9
peroxidase (EC 1.11.1.7)	*Gossypium irsutum*	7433087	2.00E-17	8
glycosyl hydrolase family 19 (chitinase)	*Arabidopsis thaliana*	15228911	1.10E-37	8
lipid transfer protein isoform 4	*Vitis vinifera*	28194086	3.00E-16	7
unknown	*Arabidopsis thaliana*	21553375	1.10E-22	7
DP-E2F-related protein 1	*Arabidopsis thaliana*	22331664	1.10E-48	7
unknown protein	*Arabidopsis thaliana*	28393189	7.10E-70	6
senescence-associated protein	*Pisum sativum*	13359451	3.10E-42	6
mannose-binding lectin	*Cymbidium hybrid*	2144226	2.10E-21	6

(Continued)

Table 11.1. *(Continued)*

Gene Annotation	Reference Organism	GI Number	E Value	ESTs
sucrose synthase	*Oncidium*	22347630	1.10E-95	6
peroxidase (EC 1.11.1.7) 2, cationic	*Glycine max*	7433098	1.10E-90	5
GDSL-motif lipase/ hydrolase protein	*Arabidopsis thaliana*	15228189	3.10E-90	5
pollen allergen-like protein	*Arabidopsis thaliana*	21593946	6.10E-20	5
No hits found				5
peroxidase	*Glycine max*	5002234	7.00E-18	5

(From Tan *et al.*, *Biotech Letts* (2005) **27:** 1517–1528. The authors wish to thank *Biotechnology Letters* for providing the figures and tables cited in the text.)

were annotated with a gene name and could be analyzed further. From Fig. 11.1A, it is obvious the length of subtractive ESTs had a linear relationship with the number of ESTs that could be annotated. The longer the sequence, the greater the chance of being annotated and vice versa.

Using the GO classification system, only 155 ESTs (25.2% of 614 annotated ESTs) were matched, protein sequences with special GI numbers which had been previously classified on three different GO trees (cell component, molecular function, and cellular process) (Fig. 11.2). This represented three different points of view of classified genes. There were 104 classified ESTs belonging to cellular components, 29 ESTs belonging to extracellular components, and 19 ESTs to unlocalized components. According to the molecular function tree, the majority (85 ESTs) were classified as enzymes, 44 ESTs as binding proteins, and 20 as transporters. A total of 130 ESTs were involved in physiological processes, and 42 were involved in cellular processes.

11.4 Northern Blot Check of the Subtractive Efficiency of the Subtractive ESTs Dataset

From the ESTs clones, 16 sequences, including eight abundant transcripts, were selected as probes for Northern blot (Fig. 11.3). Sodium/dicarboxylate

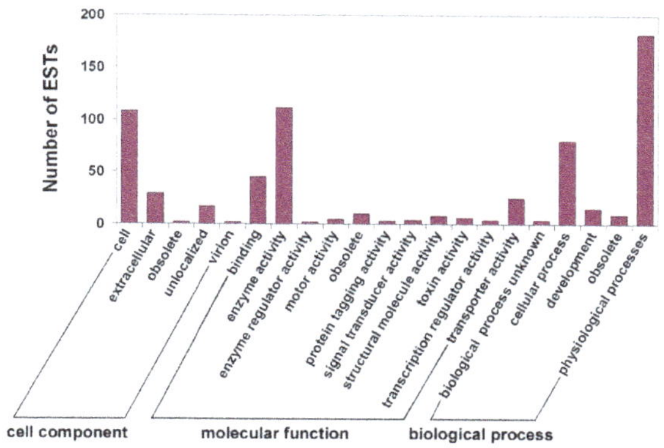

Fig. 11.2. Classification of subtractive ESTs with 3 GO trees.

(From Tan *et al.*, *Biotech Letts* (2005) **27:** 1517–1528. The authors wish to thank *Biotechnology Letters* for providing the figures and tables cited in the text.)

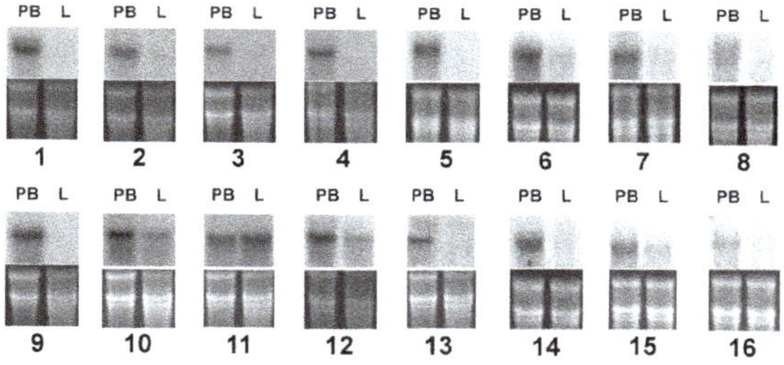

1 sodium/dicarboxylate cotransporter [*Arabidopsis thaliana*]
2 mannose-binding lectin [*Cymbidium hybrid*] 1
3 mannose-binding lectin [*Cymbidium hybrid*] 2
4 BURP domain protein [*Oryza sativa*]
5 peroxidase [*Gossypium irsutum*]
6 sucrose synthase [*Oncidium*]
7 GDP-mannose pyrophosphorylase [*Oryza sativa*]
8 mannose-6-phosephate isomerase [*Arabidopsis thaliana*]
9 proline-rich-like protease inhibitor [*Asparagus officinalis*]
10 glycine_rich RNA binding protein [*Oryza sativa*]
11 AP2 domain transcription factor [*Arabidopsis thaliana*]
12 leucine_rich receptor-related protein kinase [*Arabidopsis thaliana*]
13 Na+/H+ antiporter isoform 2 [*Lycopersicon esculentum*]
14 Invertase [*Zea Mays*]
15 granule-bound starch synthase [*Pisum sativum*]
16 pectate lyase [*Arabidopsis thaliana*]

Fig. 11.3. Northern blot check of 10 subtractive ESTs. PB, pseudobulb; L, leaf. 1–8 were subtractive ESTs with redundancy, and 9–16 were without redundancy.

(From Tan *et al.*, *Biotech Letts* (2005) **27:** 1517–1528. The authors wish to thank *Biotechnology Letters* for providing the figures and tables cited in the text.)

cotransporter, BURP domain protein (dehydration-responsive protein RD22), peroxidase, mannose-6-phosephate isomerase, proline-rich-like protease inhibitor, Na+/H+ antiporter isoform 2, invertase, and pectate lyase were highly expressed in the pseudobulbs at the initiation of inflorescence, but they exhibited almost no expression in its upper leaf. The expression patterns of mannose binding lectin, sucrose synthase, GDP-mannose pyrophosphorylase, and granule-bound starch synthase were similar. Glycine-rich RNA binding protein and leucine-rich receptor-related protein kinase genes were highly expressed in the pseudobulb and slightly in leaves at the same time. Only AP2 domain transcription factor gene displayed the opposite quality, the expression level in pseudobulbs was high, but that in leaves seemed higher. Obviously, abundant ESTs had more specific expression profiles. Significantly, the results indicated that subtractive ESTs dataset showed pseudobulb-specific genes expressed at the initiation of the inflorescence. Thus, these Northern data demonstrated that the EST subtraction was very precise and reliable.

11.5 Alignment of Subtractive ESTs with Known Sucrose Synthase and Peroxidase Genes

Sucrose synthase was the only known *Oncidium* gene as a reference sequence for ESTs annotation. Using blastn, three EST contigs and three singletons were aligned with 2520 bp, a complete CDS. They were aligned on four different regions of the gene and resulted in three clusters and one single (Table 11.3). This is because *Rsa*I digested a complete cDNA into several pieces. It also hinted that perhaps two gene family members were expressed at that time. Obviously, this example helped explain why several clusters or singles had the same annotation (Fig. 11.4A).

Another explanation may lie in the large gene family. All ESTs annotated by the peroxidase gene were aligned with a reference CDS of *Gossypium irsutum*. There were nine overlapped sequences, including eight contigs and one singleton, in the middle of the genes (Fig. 11.4B). Their homologous sequences were 205 pb. Based on their single nucleotide polymorphism, a phylogenetic tree was constructed with clustal W (Fig. 11.5). It showed three main groups on the tree. In Table 11.1, five annotations refer to peroxidase genes. Two annotations were sequence

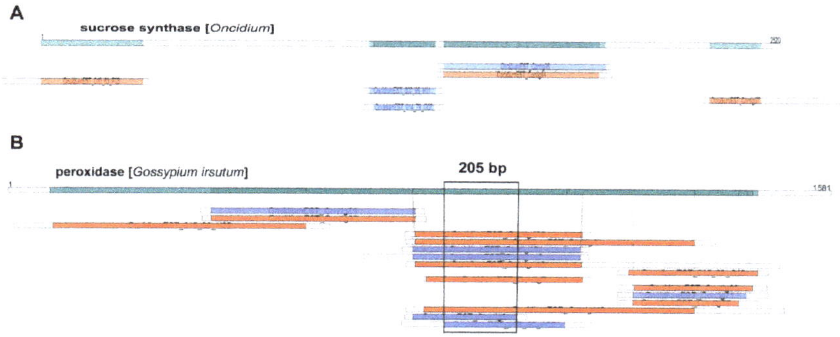

Fig. 11.4. Statement of subtractive ESTs with same annotations. (**A**) Subtractive ESTs were aligned with a known sucrose gene of *Oncidium*. (**B**) Subtractive ESTs were aligned with a known peroxidase gene of *Gossypium irsutum*.

(From Tan *et al.*, *Biotech Letts* (2005) **27:** 1517–1528. The authors wish to thank *Biotechnology Letters* for providing the figures and tables cited in the text.)

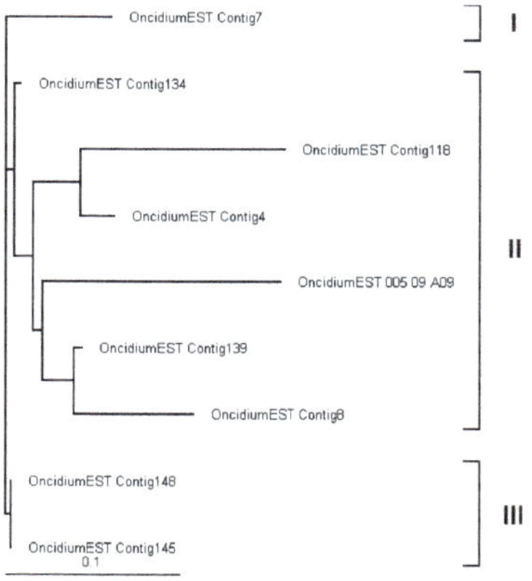

Fig. 11.5. Phylogenetic analysis of peroxidase genes expressed in pseudobulb according to sequences of overlapped 120 bp of subtractive EST contigs.

(From Tan *et al.*, *Biotech Letts* (2005) **27:** 1517–1528. The authors wish to thank *Biotechnology Letters* for providing the figures and tables cited in the text.)

consensuses belonging to the front and rear part of the gene, respectively. The other three annotations came from sequences aligned in the middle of the gene. That is to say, the peroxidase gene family was expressed in the pseudobulb, and it could be divided into three primary subfamilies.

11.6 Subtractive ESTs Relevant to Inflorescence, Carbohydrate Metabolisms, Transportation, Stress, Cell Cycle, and Regulation

According to the annotations and references about their functions, the subtractive ESTs were gathered and analyzed manually.

One cluster and seven singles (10 ESTs in total) were related to specific flower genes (Table 11.2). Abnormal inflorescence meristem 1 (AIM1) could affect inflorescence and floral development in *Arabidopsis*.[11] MADS box protein genes were expressed in different organs and mainly during floral development. DOMADS2 was expressed throughout the process of flower development and responded to osmotic changes and darkness.[12] Shaggy-like kinase was flower-specific and responsible for osmotic changes and darkness[13] inflorescence development dependent genes. The others were related to four kinds of genes within the synthesis pathway of flower pigment. Chalcone synthase[14] and chalcone-flavanone isomerase[15] were at the upper stream of the pathway. More chalcone-flavanone isomerase gene seemed to be expressed in the pseudobulb at

Table 11.2. Selected Examples of ESTs for Genes with Known or Putative Functions Related to Inflorescence

Gene Annotation	Reference Organism	GI Number	E Value	ESTs
abnormal inflorescence meristem 1 (AIM1)	*Arabidopsis thaliana*	15235527	2.00E-36	1
chalcone sythase C2	*Zea mays*	116380	6.00E-15	1
chalcone-flavanone isomerase	*Arabidopsis thaliana*	18414838	3.00E-29	3
cytochrom P450	*Arabidopsis thaliana*	21595357	2.00E-34	1
cytochrome P450 71D2	*Catharanthus roseus*	28261339	2.00E-30	1
dihydroflavonol reductase	*Arabidopsis thaliana*	18390863	1.00E-16	1
MADS box protein (DOMADS2)	*Dendrobium grex*	6467974	5.00E-24	1
shaggy-like kinase	*Ricinus communis*	1877397	8.00E-99	1

(From Tan *et al.*, *Biotech Letts* (2005) **27:** 1517–1528. The authors wish to thank *Biotechnology Letters* for providing the figures and tables cited in the text.)

the initiation of inflorescence. Two ESTs annotated with cytochrome P450 genes could also be annotated as flavanone 3′-hydroxylase or flavanone 3′-5′-hydroxylase. They catalyzed dihydrokaempferol into dihydroquercetin or dihydromyrlcetin.[16] Incorporated with UDP-glucose transferase, dihydroflavonol reductase was a downstream gene of the pathway and took part in the synthesis of anthocyanins.[16]

We found 61 subtractive ESTs annotated with genes involved in the metabolism of saccharides, including mannose, glucose, fructose, galactose, sucrose, starch, pectin, and cellulose (Table 11.3). Based on this information, we could draw a draft of carbohydrate pathways to explain what happened in pseudobulbs at the initiation of inflorescence development (data not shown).

A total of 91 ESTs were thought to have probable relationships with transportation (Table 11.4). Most of the proteins the genes encoded were localized in the kinds of membranes or on cell matrixes that help material transportation. Sodium/dicarboxylate cotransporter and GDSL-motif lipase/hydrolase were most abundant. Sodium/dicarboxylate cotransporter was a single copy gene in *Arabidopsis* and localized on vacuole membrane to transfer malate into vacuole.[17] GDSL-motif lipase/hydrolase was a lipolytic enzyme which might be related to a secretion mechanism.[18] ADP-ribosylation factor played a critical role in intracellular trafficking and maintenance of endoplasmic reticulum morphology in *Arabidopsis*.[19] Gamma-adaptin is involved in Golgi-endosome traffic, including the recruitment of accessory proteins, gamma-synergin, and Rabaptin-5.[20] Golgi-localized protein (GRIP) could maintain normal Golgi morphology and function.[21] C2 domain-containing protein was found in a large variety of membrane trafficking and signal transduction proteins and most of their biological roles have not been identified.[22] Including transporters in membranes, others interacting with cell scaffolds were also expressed, such as dynamin-like protein, F-actin capping protein, kinesin, and villin.

We identified 186 ESTs as possible stress-related genes (Table 11.5). Amazingly, 131 ESTs were peroxidase genes. In *Arabidopsis*, they are a large gene family composed of 78 members with different expression profiles in different organs.[23] Based on EST alignment, expressed peroxidase genes in pseudobulbs belong to a large family too. The others were genes induced by different biotic and abiotic stresses, including wounding, drought, and pathogens. Among them, AP2 domain transcription factor could be induced by cold, dehydration, and ABA stress,[24]

Table 11.3. Selected Examples of ESTs for Genes Related to Carbohydrate Metabolisms

Gene Annotation	Reference Organism	GI Number	E Value	ESTs
ADP-glucose pyrophosphorylase	*Arabidopsis thaliana*	30699056	4.00E-66	2
aldose 1-epimerase	*Arabidopsis thaliana*	15242099	3.00E-65	2
alpha 1,4-glucan phosphorylase L isozyme	*Oryza sativa*	13195430	1.00E-43	1
alpha-galactosidase	*Arabidopsis thaliana*	11264291	3.00E-33	1
beta-1,3 glucanase	*Oryza sativa*	20161490	7.00E-39	1
beta-fructofuranosidase 1	*Zea mays*	1352468	2.00E-31	1
beta-galactosidase	*Oryza sativa*	20514290	7.00E-63	1
beta-galactosidase	*Oryza sativa*	18461259	2.00E-49	1
beta-mannosidase	*Lycopersicon esculentum*	17226270	1.00E-103	1
cinnamyl alcohol dehydrogenase	*Populus balsamifera*	9998899	3.00E-38	1
dTDP-glucose 4-6-dehydratase	*Arabidopsis thaliana*	21594350	1.00E-16	1
epimerase/dehydratase	*Oryza sativa*	20042976	1.00E-56	3
glucosyltransferase	*Arabidopsis thaliana*	25408401	2.00E-26	1
glycogenin glucosyltransferase (EC 2.4.1.186)	*Oryza sativa*	5441877	5.00E-84	1
glycogenin glucosyltransferase (EC 2.4.1.186)	*Oryza sativa*	5441877	7.00E-93	1
glycosyl hydrolase 1	*Arabidopsis thaliana*	15220627	6.00E-36	1
granule-bound starch synthase	*Pisum sativum*	15626365	4.00E-86	1
mannose-6-phosphate isomerase	*Arabidopsis thaliana*	15232927	8.00E-50	1
mannose-6-phosphate isomerase	*Oryza sativa*	11275529	6.00E-45	1

(*Continued*)

Table 11.3. (*Continued*)

Gene Annotation	Reference Organism	GI Number	E Value	ESTs
mannosyltransferase	*Arabidopsis thaliana*	22326970	4.00E-15	1
NAD-dependent epimerase/dehydratase	*Arabidopsis thaliana*	15231926	2.00E-63	1
nucleoside-diphosphate-sugar pyrophosphorylase	*Oryza sativa*	29893646	6.00E-68	4
N-acetylglucosamine-phosphate mutase	*Arabidopsis thaliana*	30686654	8.00E-38	1
pectate lyase	*Arabidopsis thaliana*	10177179	4.00E-23	1
pectin esterase	*Oryza sativa*	20161185	1.00E-12	2
pectinesterase 1	*Lycopersicon esculentum*	6174913	1.00E-13	4
phosphoglucomutase, cytoplasmic	*Solanum tuberosum*	12585316	2.00E-84	1
phosphoglucose isomerase	*Dioscorea septemloba*	2351056	2.00E-80	1
phosphomannomutase	*Arabidopsis thaliana*	15225896	2.00E-16	2
polygalacturonase	*Pisum sativum*	13958032	3.00E-57	3
polygalacturonase	*Arabidopsis thaliana*	18412253	2.00E-22	1
ripening-related protein	*Vitis vinifera*	7406669	4.00E-79	3
starch phosphorylase	*Ipomoea batatas*	12658431	2.00E-31	1
sucrose synthase	*Oncidium*	22347630	1.00E-96	6
sucrose synthase	*Oncidium*	22347630	2.00E-27	2
sucrose synthase	*Oncidium*	22347630	1.00E-38	2
sucrose synthase	*Oncidium*	22347630	4.00E-30	1
triosephosphate isomerase, cytosolic (TIM)	*Petunia x hybrida*	1351279	4.00E-27	1

(From Tan *et al.*, *Biotech Letts* (2005) **27:** 1517–1528. The authors wish to thank *Biotechnology Letters* for providing the figures and tables cited in the text.)

and was involved in the regulation of low-temperature responsive genes in barley.[25]

A total of 23 ESTs were found to have functions in the cell cycle (Table 11.6). That is to say, the cells in pseudobulbs kept growing actively and differentiating at this stage. DP-E2F-related protein 1 and

Table 11.4. Selected Examples of ESTs for Genes Related to Transportation

Gene Annotation	Reference Organism	GI Number	E Value	ESTs
Acyl-CoA-binding protein	*Panax ginseng*	19352190	3.00E-19	2
ADP-ribosylation factor	*Glycine max*	4324967	2.00E-52	2
ADP-ribosylation factor	*Oryza sativa*	18844784	2.00E-25	1
C2 domain-containing protein	*Arabidopsis thaliana*	15239959	2.00E-36	1
C2 domain-containing protein	*Arabidopsis thaliana*	15223764	5.00E-15	1
C2 domain-containing protein	*Arabidopsis thaliana*	15217968	5.00E-37	1
dynamin like protein 2a	*Arabidopsis thaliana*	19032337	2.00E-44	1
F-actin capping protein, alpha subunit	*Oryza sativa*	23617186	2.00E-12	1
gamma-adaptin 1	*Oryza sativa*	19386749	1.00E-30	1
GDSL-like lipase/ acylhydrolase	*Oryza sativa*	29837765	1.00E-26	4
GDSL-like lipase/ acylhydrolase	*Oryza sativa*	29837765	9.00E-34	2
GDSL-motif lipase/ hydrolase	*Arabidopsis thaliana*	15228189	3.00E-91	5
GDSL-motif lipase/ hydrolase	*Arabidopsis thaliana*	21593518	2.00E-25	3
GDSL-motif lipase/ hydrolase	*Arabidopsis thaliana*	15221260	6.00E-54	2
GDSL-motif lipase/ hydrolase	*Arabidopsis thaliana*	18416824	2.00E-25	1
GDSL-motif lipase/ hydrolase	*Arabidopsis thaliana*	15224201	4.00E-12	1
golgi-localized protein (GRIP)	*Oryza sativa*	22093862	2.00E-18	1
high mobility group protein 2	*Arabidopsis thaliana*	15231065	7.00E-31	1
kinesin	*Daucus carota*	15186760	3.00E-30	2
kinesin-related protein	*Arabidopsis thaliana*	22327641	3.00E-54	3
lipid transfer protein isoform 4	*Vitis vinifera*	28194086	3.00E-16	7

(Continued)

Table 11.4. *(Continued)*

Gene Annotation	Reference Organism	GI Number	E Value	ESTs
membrane bound O-acyl transferase (MBOAT)	*Arabidopsis thaliana*	22329514	9.00E-24	1
membrane bound O-acyl transferase (MBOAT)	*Arabidopsis thaliana*	22329514	1.00E-75	1
mitochondrial carrier protein	*Arabidopsis thaliana*	15240756	5.00E-39	1
myosin heavy chain	*Arabidopsis thaliana*	18402909	5.00E-17	1
permease	*Oryza sativa*	27545049	2.00E-17	3
peroxisomal targeting signal type 1 receptor	*Arabidopsis thaliana*	15241175	3.00E-15	1
PEX14 protein	*Arabidopsis thaliana*	30697742	3.00E-15	1
plasma membrane intrinsic protein	*Oryza sativa*	22831004	2.00E-44	4
Rer1A protein (AtRer1A)	*Oryza sativa*	10945247	2.00E-37	2
Sec31p	*Oryza sativa*	22831279	4.00E-63	1
secretory carrier membrane protein	*Arabidopsis thaliana*	15222550	1.00E-30	1
signal peptidase	*Arabidopsis thaliana*	15240934	7.00E-56	1
sodium/dicarboxylate cotransporter	*Arabidopsis thaliana*	15238130	5.00E-44	24
sodium/dicarboxylate cotransporter	*Arabidopsis thaliana*	15238130	2.00E-39	3
sodium-dicarboxylate cotransporter	*Arabidopsis thaliana*	21536650	2.00E-35	1
vesicle transport v-SNARE protein	*Oryza sativa*	19571103	8.00E-55	1
villin 1 (VLN1)	*Arabidopsis thaliana*	26451417	2.00E-22	1

(From Tan *et al.*, *Biotech Letts* (2005) **27:** 1517–1528. The authors wish to thank *Biotechnology Letters* for providing the figures and tables cited in the text.)

homobox genes were most abundant among this group. The E2F/DP protein family controls cell cycle progression by acting predominantly as an activator or repressor of transcription.[26] *Arabidopsis* had more than 180 potential E2F target genes with various functions: cell cycle, transcription, stress and defense, or signaling.[27] Homeobox 20 had a

Table 11.5. Selected Examples of ESTs for Known or Putative Stress-related Genes

Gene Annotation	Reference Organism	GI Number	E Value	ESTs
acid phosphatase	*Arabidopsis thaliana*	22330531	1.00E-35	1
AP2 domain transcription factor	*Arabidopsis thaliana*	21593696	2.00E-18	1
AP2 domain transcription factor	*Arabidopsis thaliana*	21593696	5.00E-58	1
beta-N-acetylhexosaminidase	*Arabidopsis thaliana*	21537026	4.00E-50	1
biostress-resistance-related protein	*Triticum aestivum*	29409364	1.00E-61	1
bZIP DNA-binding protein	*Capsicum chinense*	4457221	3.00E-27	1
chloroplastic light-induced, drought-induced stress protein	*Solanum tuberosum*	22261807	4.00E-40	1
choline monooxygenase	*Suaeda liaotungensis*	21217447	6.00E-19	1
dehydration-induced protein	*Arabidopsis thaliana*	18411430	2.00E-68	1
DHHC-type zinc finger domain-containing protein	*Arabidopsis thaliana*	18409331	2.00E-34	1
disease resistance protein	*Arabidopsis thaliana*	15232373	9.00E-26	3
disease resistance protein (NBS-LRR class)	*Arabidopsis thaliana*	15231860	4.00E-18	1
extensin	*Populus nigra*	7484770	6.00E-42	4
farnesyltranstransferase	*Oryza sativa*	20160508	7.00E-11	1
glyceraldehyde 3-phosphate dehydrogenase, cytosolic	*Magnolia quinquepeta*	120669	2.00E-99	2
glycosyl hydrolase family 19 (chitinase)	*Arabidopsis thaliana*	15228911	1.00E-38	8
heat shock protein	*Arabidopsis thaliana*	15225377	3.00E-20	1

(*Continued*)

Table 11.5. *(Continued)*

Gene Annotation	Reference Organism	GI Number	E Value	ESTs
heat shock protein	*Arabidopsis thaliana*	15225377	1.00E-40	1
heat shock protein cognate 70	*Oryza sativa*	29124135	2.00E-42	1
heat shock protein hsc70-3 (hsc70.3)	*Arabidopsis thaliana*	15232682	7.00E-29	1
late embryogenesis abundant protein	*Arabidopsis thaliana*	15224810	2.00E-16	1
leucine rich repeat protein	*Arabidopsis thaliana*	30686169	1.00E-37	1
major intrinsic protein (MIP)	*Arabidopsis thaliana*	15236485	1.00E-81	2
Na+/H+ antiporter 2	*Lycopersicon esculentum*	15982206	4.00E-17	1
nodulin	*Oryza sativa*	11072005	9.00E-31	1
PDR-like ABC transporter	*Oryza sativa*	27368827	4.00E-35	1
peroxidase	*Glycine max*	5002234	7.00E-18	5
peroxidase (EC 1.11.1.7)	*Gossypium irsutum*	7433087	2.00E-48	92
peroxidase (EC 1.11.1.7)	*Gossypium irsutum*	7433087	2.00E-17	8
peroxidase (EC 1.11.1.7) 2, cationic	*Glycine max*	7433098	6.00E-27	21
peroxidase (EC 1.11.1.7) 2, cationic	*Glycine max*	7433098	1.00E-91	5
phosphoethanolamine methyltransferase	*Oryza sativa*	22535531	8.00E-13	1
plastid-lipid associated protein PAP/fibrillin	*Arabidopsis thaliana*	18403751	2.00E-35	1
proline rich protein 3	*Cicer arietinum*	21615411	5.00E-75	1
proline-rich protein APG isolog	*Cicer arietinum*	10638955	4.00E-16	1
proline-rich-like protein	*Asparagus officinalis*	1531756	2.00E-29	1
senescence-associated protein	*Pisum sativum*	13359451	3.00E-43	6
senescence-associated protein	*Arabidopsis thaliana*	18398417	2.00E-20	1
wound-induced protein	*Arabidopsis thaliana*	15234987	2.00E-15	3

(From Tan *et al.*, *Biotech Letts* (2005) **27:** 1517–1528. The authors wish to thank *Biotechnology Letters* for providing the figures and tables cited in the text.)

Table 11.6. Selected Examples of ESTs for Genes Related to Cell Cycle

Gene Annotation	Reference Organism	GI Number	E Value	ESTs
26S proteasome non-ATPase, regulatory subunit 6	*Oryza sativa*	20978545	8.00E-79	1
3-hydroxy-3-methylglutaryl-coenzyme A reductase 3 (HMG3.3)	*Solanum tuberosum*	11133016	5.00E-32	1
AUX1-like permease	*Arabidopsis thaliana*	5881784	2.00E-24	1
auxin efflux carrier protein	*Arabidopsis thaliana*	15239215	3.00E-50	1
biotin carboxyl carrier protein subunit	*Glycine max*	12006165	2.00E-29	1
cyclic nucleotide-regulated ion channel (CNGC9)	*Arabidopsis thaliana*	15234769	8.00E-34	1
cysteine proteinase AALP	*Arabidopsis thaliana*	23397070	6.00E-30	1
cysteine proteinase mir3 (EC 3.4.22.)	*Zea mays*	7435806	5.00E-47	1
DP-E2F-related protein 1	*Arabidopsis thaliana*	22331664	1.00E-49	7
histone deacetylase 2 isoform b	*Zea mays*	7716948	6.00E-19	1
homeobox 20	*Nicotiana tabacum*	4589882	1.00E-49	1
homeobox protein knotted-1 2 (KNAP2)	*Malus x domestica*	6016217	8.00E-46	3
homeobox-leucine zipper protein ATHB-13	*Arabidopsis thaliana*	15222452	9.00E-38	1
Homeotic protein knotted-1 (TKN1)	*Lycopersicon esculentum*	3023974	9.00E-16	1
nucleolysin	*Oryza sativa*	4680340	1.00E-15	1

(From Tan *et al.*, *Biotech Letts* (2005) **27:** 1517–1528. The authors wish to thank *Biotechnology Letters* for providing the figures and tables cited in the text.)

common motif and took part in xylem cell differentiation.[28] Homeobox-leucine zipper protein ATHB-13 was a transcription factor. It could specify the cell fate and body plan in early embryogenesis.[29]

A total of 59 ESTs were annotated with known or putative regulatory functions (Table 11.7). It seemed that the regulation of the genes

Table 11.7. Selected Examples of ESTs for Known or Putative Regulatory Functions

Gene Annotation	Reference Organism	GI Number	E Value	ESTs
adapter protein SPIKE1	*Oryza sativa*	24899400	4.00E-56	1
adenine phosphoribosyl-transferase form 2	*Oryza sativa*	29826070	8.00E-75	1
amidase	*Arabidopsis thaliana*	8163875	7.00E-31	1
amidase	*Oryza sativa*	18542894	1.00E-12	1
BURP domain protein	*Vigna unguiculata*	7106540	6.00E-12	9
c-myc binding protein	*Arabidopsis thaliana*	22325671	2.00E-12	1
cupin domain-containing protein	*Arabidopsis thaliana*	15226403	3.00E-29	2
DEAD/DEAH box helicase	*Arabidopsis thaliana*	15222526	5.00E-40	1
DEAD/DEAH box helicase	*Arabidopsis thaliana*	15219185	1.00E-26	1
DnaJ protein	*Salix gilgiana*	11277163	1.00E-106	2
DnaJ protein homolog 2	*Allium porrum*	1169382	4.00E-26	1
DNAJ-like protein	*Oryza sativa*	29367357	8.00E-33	1
elongation factor 1-alpha	*Elaeis oleifera*	18419676	8.00E-40	1
GAMYB-binding protein	*Hordeum vulgare*	27948448	7.00E-51	1
GF14 protein	*Fritillaria agrestis*	2921512	2.00E-66	1
glycine-rich RNA-binding protein	*Arabidopsis thaliana*	21553602	7.00E-21	1
HD-Zip transcription factor Athb-14	*Arabidopsis thaliana*	15226808	1.00E-96	1
helicase	*Arabidopsis thaliana*	18395518	4.00E-30	1
homeobox-leucine zipper protein ATHB-13	*Arabidopsis thaliana*	15222452	9.00E-38	1
MuDR mudrA-like protein	*Oryza sativa*	5441874	4.00E-31	1
phosphoprotein phosphatase (EC 3.1.3.16)	*Arabidopsis thaliana*	25513447	2.00E-94	1

(*Continued*)

Table 11.7. (*Continued*)

Gene Annotation	Reference Organism	GI Number	E Value	ESTs
probable protein disulfide-isomerase	*Nicotiana tabacum*	7489183	1.00E-88	1
Ras-related protein Rab11C	*Nicotiana tabacum*	3024503	5.00E-31	1
receptor-like kinase RHG4	*Glycine max*	21239384	2.00E-17	1
receptor-like protein kinase	*Arabidopsis thaliana*	7487253	6.00E-41	1
receptor-like protein kinase (EC 2.7.1)	*Oryza sativa*	7434420	3.00E-18	1
receptor-related protein kinase	*Arabidopsis thaliana*	15240720	6.00E-45	1
RNA recognition motif (RRM)-containing protein	*Arabidopsis thaliana*	22328805	8.00E-18	1
RNA-binding protein	*Oryza sativa*	18087662	2.00E-37	2
RNA-binding protein	*Mesembryanthemum crystallinum*	1076251	4.00E-26	1
serine/threonine kinase	*Arabidopsis thaliana*	25387051	2.00E-51	1
serine/threonine protein kinase	*Nicotiana tabacum*	3811293	3.00E-19	1
serine/threonine protein kinase (EC 2.7.1)	*Avena sativa*	7489361	5.00E-37	1
serine/threonine-specific protein kinase	*Arabidopsis thaliana*	25751318	2.00E-18	1
SNF2 domain/helicase domain-containing protein	*Arabidopsis thaliana*	15226870	4.00E-42	1
sphingosine kinase	*Oryza sativa*	13786462	3.00E-71	1
transcription factor LIM	*Nicotiana tabacum*	18565124	5.00E-46	2
transcription factor X1	*Oryza sativa*	6650526	2.00E-28	1
transducin/WD-40 repeat protein	*Arabidopsis thaliana*	30682603	3.00E-21	1
transfactor	*Arabidopsis thaliana*	6223653	2.00E-34	1

(*Continued*)

Table 11.7. *(Continued)*

Gene Annotation	Reference Organism	GI Number	E Value	ESTs
translational activator	*Arabidopsis thaliana*	25404492	2.00E-18	1
translational activator	*Arabidopsis thaliana*	15217742	5.00E-42	1
WD-40 repeat protein	*Arabidopsis thaliana*	30685408	5.00E-46	1
zinc finger (C3HC4-type RING finger) protein	*Arabidopsis thaliana*	15233298	9.00E-28	1
zinc finger protein	*Pisum sativum*	11288368	3.00E-77	1
zinc finger protein 5 (ZFP5)	*Arabidopsis thaliana*	21592423	5.00E-12	1
zinc-finger protein Lsd1	*Arabidopsis thaliana*	30685085	1.00E-22	1

(From Tan *et al.*, *Biotech Letts* (2005) **27:** 1517–1528. The authors wish to thank *Biotechnology Letters* for providing the figures and tables cited in the text.)

involved in the active material and energy metabolism in the pseudobulb was very complex.

11.7 Discussion

In *Oncidium* plant, the pseudobulb is a critical organ regulating plant growth. In general, when the pseudobulb grows healthily and accumulates abundant nutrition, the embedded young bud develops into inflorescence, a so-called reproductive state. However, the bud might develop into a young adventitious shoot and begin a new vegetative state. Therefore, the pseudobulb plays a functional role in regulating *Oncidium's* switch between the reproductive state and the vegetative state. In the previous study, we analyzed the carbohydrate composition and amino acid pool in the pseudobulb prior to and during inflorescence development.[10] Alternation of nutrition in the pseudobulb during floral development was also investigated.[10] To our knowledge, mannan is the principal storage polysaccharide synthesized in the pseudobulb before inflorescence initiation. It is degraded into mannose and *de novo* synthesized into starch during inflorescence development. Eventually, starch is utilized while the flower is forming.[10] Therefore, active metabolism related to carbohydrate and other types of metabolism may affect floral development. However, the genes involved in these active metabolic steps are still unknown.

Indubitably, subtractive EST data have more redundancy than EST data. Including the inevitable consequence of the nonuniform abundance of mRNA from different genes with dramatically various numbers in the cDNA pool, the bias must be further enhanced by hybridization, PCR amplification twice, ligation, and transformation. To this point, this is also the main problem in EST sequencing based on the cDNA library. It could perhaps be resolved by subtracting newly found redundant sequences from templates during stepwise sequencing. EST sequencing only from the 5′-ends of cDNAs is a rapid, economical, and amenable strategy.[6] However, for subtractive EST sequencing, not only is Rsa I digestion infeasible, it also causes shorter pieces of ESTs. Furthermore, it seemed the middle parts of genes have more chances to be sequenced than the ends (Fig. 11.4).

Thus, a large-scale analysis of gene expression related to physiological responses of the *Oncidium* pseudobulb, particularly during the early floral stage, has not been reported. Therefore, the expressed gene catalogue presented here will provide the basic information to investigate the molecular genetics basis of the *Oncidium* pseudobulb's early flowering stage by transcriptome profiling. In this small-scale subtractive EST, a conclusive picture of the cellular processes of stress-response (Table 11.5), carbohydrate metabolism (Table 11.3), and transportation (Table 11.4) was obtained. Also, the RNA gel-blot expression data showed some evidence that this EST-set is indeed enriched with such genes (Fig. 11.3), indicating the high efficiency of the cDNA subtraction strategy.

Based on the types of genes represented, the mechanism of inflorescence initiation in the early flowering stage of the *Oncidium* pseudobulb relies on precise integration of developmental processes with the carbohydrate metabolism-related response. The pseudobulb at the stage of inflorescence initiation seems like a ripening fruit for the coming flower development. First, it is in growing, including cell enlargement and differentiation. Especially, several genes involved in pectin degradation are expressing actively (Table 11.3). This may be a physiological marker for the pseudobulb, which is going to mature.[30] From the Northern results of genes related to mechanisms of saccharides at different stages of inflorescence, the ripening of pseudobulb is just in the initiation of inflorescence development (data not shown). At the same time, transportation is very active, and storage seems to be in preparation. There are two main directions of transportation: extracellular and intracellular, in which vacuoles play a predominant role. For extracellular transportation, saccharides and proteins are transferred to cell wall for synthesis. Intracellular transportation involves mainly the transfer of low molecular saccharides into vacuoles to maintain osmosis and to prepare for future storage functions.

Pseudobulb mannan is very rich and in degradation at the initiation of inflorescence. Plant mannose binding lectin is believed to play a role in the recognition of high-mannose type glycans of foreign microorganisms or plant predators.[31] However, mannose is grasped with mannose binding lectin in the pseudobulb, for it is harmful to plant cells.[32] Then the mannose can be further catalyzed into fructose and glucose. Starch is also synthesized at this stage, but almost no starch is stored in cells compared with the starch granules seen before blossoming (data not shown). Starch is perhaps also degraded into sucrose and other monohexoses for cell growth and differentiation.

At the bottom side of the pseudobulb, the little flower bud is also in differentiation and growing at the same time. Usually the bud there develops into inflorescence, but it can sometimes turn into a vegetative reproductive bud. This is a hint that the pseudobulb offers signals leading to inflorescence development unless some accident occurs. The genes involved in this function are perhaps flower-specific along with some transcriptional factors. It is no wonder so many genes are related to regulation because the pseudobulb is a very active organ leading to inflorescence at this stage. Making their functions clear is important and will require further research.

In summary, the subtractive EST approach is an efficient tool to provide an overview of gene expression profiles in the metabolically active tissue of the *Oncidium* pseudobulb. The EST data provide us with an insight into a wide range of genes. These genes represent the physiological status in the pseudobulb during the early inflorescence development. The abundant genes, e.g. peroxidase and sodium/dicarboxylate cotransporter, shown in the profile revealed some especially unexpected facts. These will someday make it possible to exploit flowering-related mechanisms for the benefit of mankind.

Acknowledgment

This work was financially supported by National Science Council, ROC under Grant NSC 91-2317-B-002-041 to Professor K. W. Yeh.

References

1. Liau CH, You SJ, Prasad V, *et al.* (2003a) Agrobacterium tumefaciens-mediated transformation of an *Oncidium* orchid. *Plant Cell Rep* **21:** 993–998.
2. Chen J, Chang W. (2000) Efficient plant regeneration through somatic embryogenesis from callus cultures of *Oncidium* (*Orchidaceae*). *Plant Sci* **160:**87–93.
3. You SJ, Liau CH, Huang HE, *et al.* (2003) Sweet pepper ferredoxin-like protein (pflp) gene as a novel selection marker for orchid transformation. *Planta* **217:**60–65.
4. Liau CH, Lu JC, Prasad V, *et al.* (2003b) The sweet pepper ferredoxin-like protein (pflp) conferred resistance against soft rot disease in *Oncidium* orchid. *Transgenic Res* **12:**329–336.
5. Hsu HF, Yang CH. (2002) An orchid (*Oncidium* Gower Ramsey) AP3-like MADS gene regulates floral formation and initiation. *Plant Cell Physiol* **43:**1198–1209.
6. Li CR, Zhang XB, Hew CS. (2003) Cloning of a sucrose-phosphate synthase gene highly expressed in flowers from the tropical epiphytic orchid *Oncidium* Goldiana. *J Exp Bot* **21:**89–91.
7. Hew CS, Yang JWH. (1994) Growth and photosynthesis of *Oncidium* Goldiana. *J Hort Sci* **69:**809–819.
8. Wadasinghe G, Hew CS. (1995) The importance of back shoots as a source of photo-assimilates for growth and flower production in *Dendrodium* cv. Jashika pink (Orchidaceae). *J Hort Sci* **70:**207–214.
9. Hew CS, Ng CKY. (1996) Changes in mineral and carbohydrate content in pseudobulbs of the C_3 epiphytic orchid hybrid *Oncidium* Goldiana at different growth stages. *Lindleyana* **11:**125–134.
10. Wang HL, Chung JD, Yeh KW. (2003) Changes of carbohydrate and free amino acid pools in current pseudobulbs of *Oncidium* Gower Ramsey during inflorescence development. *J Agric Associ China* **4:**476–488.
11. Richmond TA, Bleecker AB. (1999) A defect in beta-oxidation causes abnormal inflorescence development in *Arabidopsis*. *Plant Cell* **11:**1911–1924.
12. Yu H, Goh CJ. (2000) Identification and characterization of three orchid MADS-box genes of the AP1/AGL9 subfamily during floral transition. *Plant Physiol* **123:**1325–1336.

13. Charrier B, Champion A, Henry Y, Kreis M. (2002) Expression profiling of the whole *Arabidopsis* shaggy-like kinase multigene family by real-time reverse transcriptase-polymerase chain reaction. *Plant Physiol* **130:**577–590.
14. Wade HK, Bibikova TN, Valentine WJ, Jenkins GI. (2001) Interactions within a network of phytochrome, cryptochrome and UV-B phototransduction pathways regulate chalcone synthase gene expression in *Arabidopsis* leaf tissue. *Plant J* **25:**675–685.
15. van Tunen AJ, Koes RE, Spelt CE, *et al.* (1988) Cloning of the two chalcone flavanone isomerase genes from *Petunia hybrida*: Coordinate, light-regulated and differential expression of flavonoid genes. *EMBO J* **7:**1257–1263.
16. de Vetten N, ter Horst J, van Schaik HP, *et al.* (1999) A cytochrome b5 is required for full activity of flavonoid 3′, 5′-hydroxylase, a cytochrome P450 involved in the formation of blue flower colors. *Proc Natl Acad Sci USA* **96:**778–783.
17. Emmerlich V, Linka N, Reinhold T, *et al.* (2003) The plant homolog to the human sodium/dicarboxylic cotransporter is the vacuolar malate carrier. *Proc Natl Acad Sci USA* **100:**11122–11126.
18. Wilhelm S, Tommassen J, Jaeger KE. (1999) A novel lipolytic enzyme located in the outer membrane of *Pseudomonas aeruginosa*. *J Bacteriol* **181:**6977–6986.
19. Lee MH, Min MK, Lee YJ, *et al.* (2002) ADP-ribosylation factor 1 of *Arabidopsis* plays a critical role in intracellular trafficking and maintenance of endoplasmic reticulum morphology in *Arabidopsis*. *Plant Physiol* **129:**1507–1520.
20. Nogi T, Shiba Y, Kawasaki M, *et al.* (2002) Structural basis for the accessory protein recruitment by the gamma-adaptin ear domain. *Nat Struct Biol* **9:**527–531.
21. Ungar D, Oka T, Brittle EE, *et al.* (2002) Characterization of a mammalian Golgi-localized protein complex, COG, that is required for normal Golgi morphology and function. *J Cell Biol* **157:**405–415.
22. Ochoa WF, Corbalan-Garcia S, Eritja R, *et al.* (2002) Additional binding sites for anionic phospholipids and calcium ions in the crystal structures of complexes of the C2 domain of protein kinase calpha. *J Mol Biol* **320:**277–291.
23. Tognolli M, Penel C, Greppin H, Simon P. (2002) Analysis and expression of the class III peroxidase large gene family in *Arabidopsis thaliana*. *Gene* **288:**129–138.
24. Shen YG, Zhang WK, He SJ, *et al.* (2003) An EREBP/AP2-type protein in *Triticum aestivum* was a DRE-binding transcription factor induced by cold, dehydration and ABA stress. *Theor Appl Genet* **106:**923–930.
25. Xue GP. (2003) The DNA-binding activity of an AP2 transcriptional activator HvCBF2 involved in regulation of low-temperature responsive genes in barley is modulated by temperature. *Plant J* **33:**373–383.

26. Cayirlioglu P, Ward WO, Silver Key SC, Duronio RJ. (2003) Transcriptional repressor functions of *Drosophila* E2F1 and E2F2 cooperate to inhibit genomic DNA synthesis in ovarian follicle cells. *Mol Cell Biol* **23:**2123–2134.
27. Ramirez-Parra E, Frundt C, Gutierrez C. (2003) A genome-wide identification of E2F-regulated genes in *Arabidopsis*. *Plant J* **33:**801–811.
28. Hertzberg M, Olsson O. (1998) Molecular characterisation of a novel plant homeobox gene expressed in the maturing xylem zone of *Populus tremula* x tremuloides. *Plant J* **16:**285–295.
29. Yang JY, Chung MC, Tu CY, Leu WM. (2002) OSTF1: A HD-GL2 family homeobox gene is developmentally regulated during early embryogenesis in rice. *Plant Cell Physiol* **43:**628–638.
30. Medina-Escobar N, Cardenas J, Moyano E, *et al.* (1997) Cloning, molecular characterization and expression pattern of a strawberry ripening-specific cDNA with sequence homology to pectate lyase from higher plants. *Plant Mol Biol* **34:**867–877.
31. Chai Y, Pang Y, Liao Z, *et al.* (2003) Molecular cloning and characterization of a mannose-binding lectin gene from *Crinum asiaticum*. *J Plant Physiol* **160:**913–920.
32. Stein JC, Hansen G. (1999) Mannose induces an endonuclease responsible for DNA laddening in plant cells. *Plant Physiol* **121:**71–80.
33. Tan J, Wang HL, Yeh KW. (2005) Analysis of organ-specific, expressed genes in *Oncidium* orchid by substractive expressed sequence tags library. *Biotech Lett* **27:**1517–1528.

Chapter 12

Application of Virus-induced Gene Silencing Technology in Gene Functional Validation of Orchids

Hsiang-Chia Lu[†], *Hong-Hwa Chen*[‡] *and Hsin-Hung Yeh*[*,†]

The largest family of angiosperms, *Orchidaceae*, has diverse, specialized pollination and ecological strategies and provides a rich subject for investigating evolutionary relationships and developmental biology. However, the study of these non-model organisms may be hindered by challenges, such as their large genome size, low transformation efficiency, long regeneration time, and long life-cycle. To overcome these obstacles, we first developed vectors with the use of a symptomless *Cymbidium mosaic virus*, which infects most orchids, then combined simple physiological controls and virus-induced gene silencing (VIGS) for validation of gene function in orchids. The success of our strategies was verified by our functional validation of floral identity gene(s) in the tetraploid *Phalaenopsis* orchids, which have an unusually long life-cycle (two years from sowing to flowering). We could knock down the RNA level of either a specific *Phalaenopsis* floral identity gene or a family of genes congruently in two months. Functional analysis of orchid genes could become easier and profit from the VIGS approach.

*Corresponding author.

[†]*Department of Plant Pathology and Microbiology, National Taiwan University, Taipei, Taiwan.*

[‡]*Department of Life Sciences, National Cheng Kung University, Tainan, Taiwan.*

12.1 Introduction

Orchids, plants with elegant and beautiful flowers, have attracted humans for centuries. Known for their diversity, they form the largest family of angiosperms, *Orchidaceae*.[1] Orchids have evolved specialized and astonishing pollination and ecological strategies for survival. Therefore, they are a rich source for investigating evolutionary relationships and developmental biology. Although evolutionary developmental biology (Evo-Devo) has become an important topic in every field of biology today,[2–4] the study of orchids has been hindered by several challenges, including their large genome size, long life-cycle, and low transformation efficiency. In fact, orchids are the opposite of model plants in every trait studied by geneticists and molecular biologists. For example, the most widely grown orchid, *Phalaenopsis* spp., has a large genome size, ranging from 1×10^9 to 6×10^9 bp/1C,[5,6] and several important commercial cultivars are multiploid. The life-cycle of *Phalaenopsis* spp. naturally takes about 2 to 3 years from the vegetation to the reproductive phases. Therefore, applications of commonly used genetic and molecular approaches to the study of orchids is difficult. However, the development of several new research techniques has led to increased investigation of model plants and numerous exciting findings. From these developments, we can select suitable tools to solve the problems faced in the study of orchids.

12.2 Approaches Used for Gene Functional Analysis

Forward and reverse genetics is commonly used in the elucidation of the function of a gene or genes. Forward genetics tries to identify the genetic source of a trait of interest, beginning with random mutagenesis of the whole genome to follow the selection of the trait and subsequently identifying the mutated gene or genes related to the trait. Reverse genetics, however, aims to disrupt the expression of genes to search for a phenotype and usually begins with cloned DNA sequences, with or without prior knowledge about the DNA. In plants, disrupting the expression of a gene of interest is difficult owing to low homologous recombination frequencies. Alternatively, plants transformed with antisense and/or sense genes that can produce dsRNA from an inverted-repeat of a target gene may disrupt the RNA or block the translation of the gene.[7]

Forward genetics is not practical for functional analysis of plants, such as orchids because of their long life-cycle and lack of high-resolution genetic maps. Reverse genetics to find genes is problematic as well because of difficulties in plant transformation and long regeneration time, especially for genes involved in reproductive stages, the last stages of the life-cycle and so important to orchid study.

In addition to genetic transformation, a recently developed approach to functional analysis of plant genes involves loss-of-function mediated by virus-induced gene silencing (VIGS). Since the early success of VIGS with *Tobacco mosaic virus* (TMV)- and *Potato virus X* (PVX)-based vectors in silencing the phytoene desaturase gene (PDS) in *Nicotiana benthamiana*,[8,9] the use of virus vectors to induce plant gene silencing has been efficient and reliable in many systems, such as tomato (*Lycopersicum esculentum*), *Arabidopsis* spp., and *N. benthamiana*, some *Solanum* species; a legume species, *Pisum sativum*; soybean (*Glycin* spp.); and *Papaver somniferum*.[10–18] Among the monocots, only a *Barley strip mosaic virus*-derived vector has been developed for VIGS.[19,20]

VIGS has seen success in diverse fields of plant science. For example, *Tobacco rattle virus* (TRV)-based VIGS vectors have been applied to the study of floral development,[21] plant resistance against pathogens,[22–24] and fruit ripening.[25] A PVX-based vector has been applied in high throughput analysis to screen 4,992 cDNAs derived from *N. benthamiana*. Discovered were genes involved in signal transduction pathways of *Pto*-mediated resistance to a pathogenic bacteria, *Pseudomomas syringae*.[26]

VIGS has advantages over other approaches, such as transforming plants with antisense and/or sense genes that can produce dsRNA of target genes. The technique can be induced within a month after inoculation in every plant tested so far. In addition, it can be applied at different stages of plant development, as long as viruses can infect the plant tissues. Therefore, it prevents suppression of some essential genes required for seed maturation, a common problem with mutagenesis screening. For the detailed mechanism and advantages of VIGS, see Refs. 11, 12, 27 and 28. Therefore, VIGS offers an opportunity for functional analysis of genes in plants with long life-cycle, such as orchids.

12.3 Viral Vectors Suitable for Gene Functional Analysis of Orchids

Among the currently constructed viral vectors, only TRV can infect orchids.[29] However, despite the successful application of the TRV vector in the systems described above, its application in orchids has not been reported.[10–18] Since viral isolates may differ in their biological nature in terms of symptoms and host range, we do not know if the currently available TRV vectors are suitable for orchid gene functional analysis. For application of VIGS, the desired viral vectors must not induce symptoms in plants, so as not to complicate the functional interpretation of the gene of interest. The currently available TRV vectors cause no visible symptoms in tested plants. However, whether the TRV vectors can infect orchids and induce symptoms or not remains to be resolved. Another important issue is biological safety of the vectors. In countries, such as Taiwan, infection with TRV has not been reported in any plants. Therefore, application of TRV vectors must not violate local biosafety legislation and/or potential release of the viruses outside the tested plants.

Alternatively, orchid viruses belonging to the same genus of the currently successful VIGS vector are potential resources for developing VIGS vectors. Examples are the successful PVX vector and *Cymbidium mosaic virus* (CymMV), belonging to *Potexvirus*, and TMV and *Odontoglossum ringspot tobamovirus* (ORSV), belonging to *Tobamovirus*.

Recently, we screened and selected a symptomless CymMV, one of the most prevalent orchid viruses, to develop a VIGS vector (Fig. 12.1).[30] The vector could induce gene silencing by knocking down the RNA level of PDS in leaves and that of a specific floral identity gene or, congruently, floral identity family genes in *Phalaenopsis* in less than two months. The following describes our novel use of VIGS in functional analysis of floral identity genes in *Phalaenopsis*.

12.4 Applicability of Vigs in the Study of Gene Function in Leaves of Orchids

To monitor VIGS efficacy, we inserted a *Phalaenopsis amabilis* var. *formosa* PDS fragment into our CymMV vector (pCymmV-pro60) and inoculated *P. amabilis* var. *formosa* with the vector. Real-time RT-PCR was

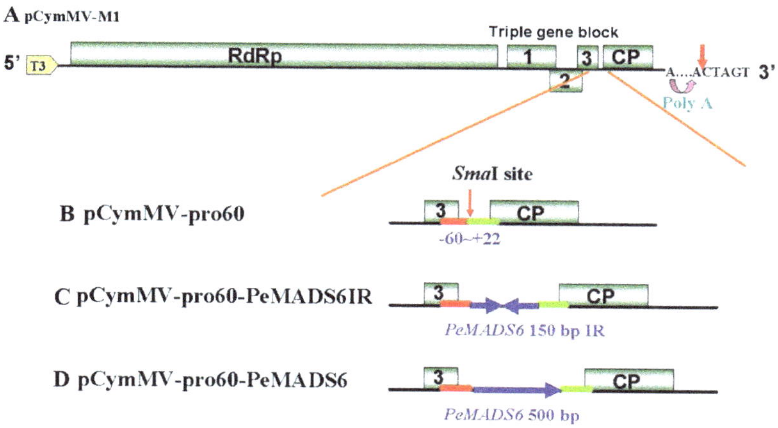

Fig. 12.1. Schematic representation of CymMV cDNA infectious clone and derivative vectors. Schematic representation of CymMV infectious clone (**A**), and derivative vectors (**B–D**). Rectangles represent ORFs encoded by CymMV genomic RNA. RNA-dependent RNA polymerase (RDRP), triple gene block 1, 2, and 3, capsid protein (CP) are indicated. The T3 promoter immediately adjacent to the most 5′-end and a poly(A) tail (25 adenosine) at the most 3′-end are indicated. The *Spe*I restriction enzyme digestion site is indicated by a red arrow. The numbers below the red lines indicate the selected region of CP subgenomic promoter, and plus or minus corresponds to the upstream and downstream CP translation start codon, respectively. The light green line indicates the duplicated CP subgenomic promoter. Head to head arrows on (**C**) indicate that the selected 150 bp of *PeMADS6* was cloned as an inverted repeat. The direct arrow on (**D**) indicates that the *PeMADS6* conserved region was cloned directly.

used to monitor the knockdown of PDS in leaves beginning at three weeks after inoculation (approximately three weeks are needed for the CymMV to establish systemic infection). The RNA level of PDS was gradually reduced to 54% in plants at eight weeks postinoculation, the level increasing thereafter; thus, gene-silencing efficacy was reduced after eight weeks. Our results showed that the CymMV vector can induce gene silencing in leaves of orchids and are consistent with a previous report indicating that VIGS efficacy is progressively reduced over time.[21] However, the result also revealed the limitation of the application of VIGS in leaves of *Phalaenopsis* orchids. PDS is a visual marker commonly used for VIGS because the bleaching phenotype is easily observed,[8,9,21] but only on newly emerging leaves, not fully expanded leaves. Because *Phalaenopsis* is a Crassulacean acid metabolism (CAM) plant with very slow growth rate (more than six months to generate a new leaf), no new leaves were able to generate before the VIGS efficacy was reduced. Therefore, no visible phenotype was observed on the plants silenced with PDS. Although the CymMV vector could induce gene silencing in the leaves of orchids, a visible phenotype could not be easily observed on those leaves. However, since gene silencing was induced in leaves, the transcriptome was changed, and an invisible phenotype might be measured or analyzed by biochemical or molecular biology approaches.

12.5 Strategies for Functional Validation of Genes Involved in the Reproductive Stage of Orchids

For functional validation of orchid genes, the most intriguing and challenging studies are those analyzing genes involved in the reproductive stages, intriguing because of the diversity of orchid flowers and challenging because in orchids, the transition from vegetative to reproductive stages usually takes about 2 to 3 years. In addition, orchids usually bloom only once a year, which limits the time for conducting experiments.

To overcome such obstacles in the study of orchids, we first learned from commercial orchid growers. Commercial orchid growers usually keep their plants under low temperature controls (25°C/day and 20°C/night) with appropriate humidity and fertilization to induce stalks for flowering.[31] With this process, orchid growers have flowers to sell year-round and we have more flowers for study.

To test whether the CymMV vector (pCymMV-pro60) could induce gene silencing in all floral organs, we selected a B-class MADS-box family gene, *PeMADS6*, which is transcribed in all flower organs.[32] Since MADS-box family genes are less conserved in their 3′-terminus, we selected a stretch of 150 nucleotides (nts) in this region and inserted it to the CymMV vector. We directly inoculated the CymMV vector containing the 150 nts of *PeMADS6* [pCymMV-pro60-PeMADS6IR (Fig. 12.1B)] into emerging stalks with six nodes (about 8 cm) of *P. amabilis* var. *formosa*. Approximately six weeks later, the flowers blossomed. Compared with mock-inoculated plants, the plants inoculated with pCymMV-pro60-PeMADS6IR showed reduced *PeMADS6* RNA levels in sepals, petals, lips, and columns to 63 ± 2%, 33 ± 3%, 23 ± 5% and 33 ± 2%, respectively, as measured by real-time RT-PCR.[30] However, the plants inoculated with the pCymMV-pro60 vector showed a *PeMADS6* RNA level similar to that of mock-inoculated plants. We also analyzed the RNA level of *PeMADS1* and *PeMADS3*, which belong to C- and B-class-like MADS-box genes, in the plants inoculated with pCymMV-pro60-PeMADS6IR. No obvious transcriptional changes in *PeMADS1* and *PeMADS3* were detected among the plants inoculated with the buffer, pCymMV-pro60, or pCymMV-pro60-PeMADS6IR. Thus, only *PeMADS6* was silenced in pCymMV-pro60-PeMADS6IR-inoculated plants.

To confirm that the reduced expression of *PeMADS6* was due to the presence of RNAi mediated by VIGS, we purified small-molecular-weight RNA from inoculated plants and performed Northern blot hybridization, with the CymMV CP gene or *PeMADS6* used as probes. CymMV could induce 21-nt siRNA, because the CymMV CP probe could detect 21-nt siRNA in both CymMV-pro60- and pCymMV-pro60-PeMADS6IR-inoculated plants. However, the *PeMADS6* 21-nt siRNA was detected only in plants inoculated with pCymMV-pro60-PeMADS6IR (the *PeMADS6* region used for probes is different from that inserted in pCymMV-pro60-PeMADS6). No siRNA was detected in the mock-inoculated plants using of both probes. Thus, the generation of *PeMADS6* siRNA is specific, and the reduction of the *PeMADS6* RNA level is through the gene-silencing mechanism.[30]

However, we observed no visible morphologic changes in *PeMADS6*-silenced plants, perhaps because the knockdown level induced by VIGS was enough to induce flower morphologic change. The expression of MADS-box genes is dose-dependent, and such genes may require complete silencing to produce a phenotype.[33–35]

12.6 Simultaneous Knockdown of MADS-Box Family Genes

Family genes with redundant functions are not easily targeted by genetic knockout assay. In addition, a vector that can easily induce a visible phenotype in orchids during VIGS will be desirable for further research. Therefore, we inserted a 500-bp DNA fragment of a *PeMADS6* conserved region of the MADS-box family of genes into pCymMV-pro60 [pCymMV-pro60-PeMADS6 (Fig. 12.1C)]. We expected that several MADS-box family genes would be affected, with consequent prominent morphologic changes. In plants inoculated with pCymMV-pro60-PeMADS6, the flower blossomed, but streaks or patches of greenish tissue appeared in the sepals, petals, and lips (Fig. 12.2). Interestingly, some of the inoculated plants initially produced flower buds on the lower stalks, but the buds could not blossom (Fig. 12.2A). The flower buds on the upper stalk could blossom to some extent, but streaks or patches of greenish tissue were observed in the sepals, petals, and lips. We dissected some initial flower buds that turned yellow (an indication that these buds would eventually abort) and found fully formed sepals, petals, lips, and columns within the buds and morphology similar to that of green buds of healthy plants. These results suggest that the reduced transcript level of MADS-box family genes still allowed the flower to develop normally at first, but not enough for the flower to further develop and blossom. Real-time RT-PCR revealed that *PeMADS6* and two randomly selected C- and B-class-like MADS-box genes, *PeMADS1* and *PeMADS3*, were all silenced in plants inoculated with pCymMV-pro60-PeMADS6. Especially, for the three analyzed genes, more than 85% of the transcript level was reduced in the initial flower buds.[30] In addition, the transcript level of all the analyzed genes was similar, and no phenotype described above was observed in either the mock- or pCymMV-pro60-inoculated plants.

The MADS-box family gene-silenced plants were unable to blossom or showed a greenish structure in the sepals and lips. This result was consistent with the prediction of the well-known ABC model, which suggests that A-, B-, and C-class-like MADS-box genes act cooperatively to control floral identity.[36,37]

We performed experiments in two different orchid varieties, *P. amabilis* and *P.* Sogo Musadium and produced similar results. Both varieties are tetraploid and, in general, loss-of-function assays are not easy to perform in plants with a multiploid genome or when the genes of interest have multiple copies. With loss-of-function assays, such as T-DNA insertion or transposon tagging, simultaneously targeting all genes with functional redundancy is difficult. In contrast, VIGS can knock down the RNA level after RNA transcription, regardless of the RNAs transcribed from genes in different genome locations and, thus can silence all genes simultaneously.

Fig. 12.2. Phenotype on MADS-box gene-silenced plants *P.* Sogo Musadium infected with pCymMV-pro60 (**B**, **D**) and pCymMV-CP60- PeMADS6 (**A**, **C**, **E**). The arrow on (**A**) indicates an aborted flower bud. Below this node, flower buds were all aborted, and flower buds produced beyond this node blossomed normally. The arrows on (**C**) indicate the greenish streaks in the sepal and petals. The arrows on (**E**) indicate the greenish patches in the lip.

12.7 Perspective

We demonstrated that our newly constructed CymMV-based vector is suitable for analyzing genes involved in floral morphogenesis of *Phalaenopsis* spp. As CymMV has a wide host range among species belonging to *Orchidaceae*, including *Phalaenopsis*, *Cymbidium*, *Cattleya*, *Dendrobium*, *Epidendrum*, *Laelia*, *Laeliocattleya*, *Oncidium*, *Zygopetalum*, *Vanilla*, and *Vinda*,[38,39] the developed vectors will contribute well to functional genomic studies of orchids. With the advent of genomics and bioinformatics, the past few years have seen great strides in the amount of gene information available and development of tools for their analysis of orchids.[40–42] Here, we have demonstrated that VIGS can be used for the functional validation of genes in orchids, difficult plants to study by the traditional genetic approaches because of their large genome size, low transformation efficiency, long regeneration time, and long life-cycle. The study of orchids will flourish and profit from approaches such as VIGS.

Acknowledgments

We thank Dr. Wen-Hui Chen (Department of Life Sciences, National Science Council, National University of Kaohsiung, Kaohsiung, Taiwan) for orchid variety consulting and Dr. Hong-Ji Su for helping to collect virus isolates. We also thank the National Science Council and Council of Agriculture in Taiwan for supporting the research grants in HHY's laboratory.

References

1. Dressler RL. (1993) *Phylogeny and Classification of the Orchid Family*. Dioscorides Press, Portland, Oregon.
2. Pennisi E. (2004) RNAi takes Evo-devo world by storm. *Science* **304:**384.
3. Babbitt C, Giorgianni M, Price A. (2002) Evo-devo comes into focus. *BioEssay* **24:**677–679.
4. Kramer EM, Hall JC. (2005) Evolutionary dynamics of genes controlling floral development. *Curr Opin Plant Bio* **8:**13–18.
5. Chen WH, Kao YY, Lin TY, *et al.* (2001) Genomic study of *Phalaenopsis* orchid. *Proceedings of APOC7*, Nagoya, Japan.

6. Lin S, Lee HC, Chen WH, *et al.* (2001) Nuclear DNA contents of *Phalaenopsis sp.* and *Doritis pulcherrima*. *J Am Soc Hortic Sci* **126:**195–199.
7. Kusaba M. (2004) RNA interference in crop plants. *Curr Opin Biotechnol* **15:**139–143.
8. Kumagai MH, Donson J, della-Cioppa G, *et al.* (1995) Cytoplasmic inhibition of carotenoid biosynthesis with virus-derived RNA. *Proc Natl Acad Sci USA* **92:**1679–1683.
9. Ruiz MT, Voinnet O, Baulcombe DC. (1998) Initiation and maintenance of virus-induced gene silencing. *Plant Cell* **10:**937–946.
10. Wang MB, Waterhouse PM. (2002) Application of gene silencing in plants. *Curr Opin Plant Biol* **5:**146–150.
11. Benedito VA, Visser PB, Angenent GC, Krens FA. (2004) The potential of virus-induced gene silencing for speeding up functional characterization of plant genes. *Genet Mol Res* **3:**323–341.
12. Burch-Smith TM, Anderson JC, Martin GB, Dinesh-Kumar SP. (2004) Applications and advantages of virus-induced gene silencing for gene function studies in plants. *Plant J* **39:**734–746.
13. Robertson D. (2004) VIGS vectors for gene silencing: Many targets, many tools. *Annu Rev Plant Biol* **55:**495–519.
14. Brigneti G, Martin-Hernandez AM, Jin H, *et al.* (2004) Virus-induced gene silencing in *Solanum* species. *Plant J* **39:**264–272.
15. Constantin GD, Krath BN, MacFarlane SA, *et al.* (2004) Virus-induced gene silencing as a tool for functional genomics in a legume species. *Plant J* **40:**622–631.
16. Fofana IB, Sangare A, Collier R, *et al.* (2004) A geminivirus-induced gene silencing system for gene function validation in cassava. *Plant Mol Biol* **56:**613–624.
17. Hileman LC, Drea S, Martino G, *et al.* (2005) Virus-induced gene silencing is an effective tool for assaying gene function in the basal eudicot species *Papaver somniferum* (opium poppy). *Plant J* **44:**334–341.
18. Zhang C, Ghabrial SA. (2006) Development of Bean pod mottle virus-based vectors for stable protein expression and sequence-specific virus-induced gene silencing in soybean. *Virology* **344:**401–411.
19. Holzberg S, Brosio P, Gross C, Pogue GP. (2002) Barley stripe mosaic virus-induced gene silencing in a monocot plant. *Plant J* **30:**315–327.
20. Scofield SR, Huang L, Brandt AS, Gill BS. (2005) Development of a virus-induced gene-silencing system for hexaploid wheat and its use in functional analysis of the Lr21-mediated leaf rust resistance pathway. *Plant Physiol* **138:**2165–2173.
21. Ratcliff F, Martin-Hernandez AM, Baulcombe DC. (2001) Technical Advance. Tobacco rattle virus as a vector for analysis of gene function by silencing. *Plant J* **25:**237–245.

22. Peart JR, Cook G, Feys BJ, *et al.* (2002) An *EDS1* orthologue is required for N-mediated resistance against tobacco mosaic virus. *Plant J* **29:**569–579.
23. Liu Y, Schiff M, Marathe R, Dinesh-Kumar SP. (2002) Tobacco *Rar1*, *EDS1* and *NPR1/NIM1* like genes are required for N-mediated resistance to tobacco mosaic virus. *Plant J* **30:**415–429.
24. Sharma PC, Ito A, Shimizu T, *et al.* (2003) Virus-induced silencing of *WIPK* and *SIPK* genes reduces resistance to a bacterial pathogen, but has no effect on the INF1-induced hypersensitive response (HR) in *Nicotiana benthamiana*. *Mol Gen Genomics* **269:**583–591.
25. Fu DQ, Zhu BZ, Zhu HL, *et al.* (2005) Virus-induced gene silencing in tomato fruit. *Plant J* **43:**299–308.
26. Lu R, Malcuit I, Moffett P, *et al.* (2003) High throughput virus-induced gene silencing implicates heat shock protein 90 in plant disease resistance. *The EMBO J* **22:**5690–5699.
27. Marathe R, Anandalakshmi R, Smith TH, *et al.* (2000) RNA viruses as inducers, suppressors and targets of post-transcriptional gene silencing. *Plant Mol Biol* **43:**295–306.
28. Burch-Smith TM, Schiff M, Liu Y, Dinesh-Kumar SP. (2006) Efficient virus induced gene silencing in *Arabidopsis thaliana*. *Plant Physiol* **142:**21–27.
29. Korotieieva HV, Polishchuk VP. (2004) Viruses of orchids of the natural flora in Ukraine. *Mikrobiol Z* **66:**74–80.
30. Lu HC, Chen HH, Tsai WC, *et al.* (2006) Strategies for functional validation of genes involved in reproductive stages of orchids. (Accepted)
31. Gorden B. (1998) *Phalaenopsis* flower induction (or, how to make them bloom). *Am Orchid Soc Bull* **9:**908–910.
32. Tsai WC, Lee PF, Chen HI, *et al.* (2005) *PeMADS6*, a *GLOBOSA/PISTILLATA*-like gene in *Phalaenopsis equestris* involved in petaloid formation, and correlated with flower longevity and ovary development. *Plant Cell Physiol* **46:**1125–1139.
33. Scortecci KC, Michaels SD, Amasino RM. (2001) Identification of a MADS-box gene, FLOWERING LOCUS M, that represses flowering. *Plant J* **26:**229–236.
34. Yu H, Xu Y, Tan EL, Kumar PP. (2002) AGAMOUS-LIKE 24, a dosage-dependent mediator of the flowering signals. *Proc Natl Acad Sci USA* **99:**16336–16341.
35. Zachgo S, Silva Ede A, Motte P, *et al.* (1995) Functional analysis of the Antirrhinum floral homeotic *DEFICIENS* gene *in vivo* and *in vitro* by using a temperature-sensitive mutant. *Development* **121:**2861–2875.
36. Bowman JL, Smyth DR, Meyerowitz EM. (1991) Genetic interactions among floral homeotic genes of *Arabidopsis*. *Development* **112:**1–20.
37. Coen ES, Meyerowitz EM. (1991) The war of the whorls: Genetic interactions controlling flower development. *Nature* **353:**31–37.

38. Zettler FW, KO NJ, Wisler GC, *et al*. (1990) Viruses of orchids and their control. *Plant Dis* **74:**621–626.
39. Hu SJ, Ferreira S, Wang M, Xu MQ. (1993) Detection of *Cymbidium mosaic virus*, *Odontoglossum ringspot virus*, *Tomato spotted wilt virus*, and Potyviruses infecting Orchids in Hawaii. *Plant Dis* **77:**464–468.
40. Tsai WC, Hsiao YY, Lee SH, *et al*. (2006) Expression analysis of the ESTs derived from the flower buds of *Phalaenopsis equestris*. *Plant Sci* **170:** 426–432.
41. Wang HC, Kuo HC, Chen HH, *et al*. (2005) KSPF: Using gene sequence patterns and data mining for biological knowledge management. *Expert Syst Appl* **28:**537–545.
42. Hsiao YY, Tsai WC, Kuoh CS, *et al*. (2006) Comparison of transcripts in *Phalaenopsis bellina* and *Phalaenopsis equestris* (Orchidaceae) flowers to deduce monoterpene biosynthesis pathway. *BMC Plant Biol*. (Accepted)

Chapter 13

Genetic Transformation as a Tool for Improvement of Orchids

Sanjaya[†] *and Ming-Tsair Chan*[*,†]

Orchids are primarily grown for their large, long-lasting, and fascinating flowers; thus, the improvement of quality attributes such as flower color, longevity, shape, architecture, biotic and abiotic stress tolerance, and creation of novel variations are important economic goals for floriculturists across the world. Recent advances in genetic engineering and molecular biology techniques augmented with gene transformation could help growers to meet the demand of the orchid industry in the new century. Presented is an overview of *Agrobacterium*-mediated and particle bombardment or direct gene transformation in orchids and the essential factors involved in the technology systems. Available methods for the transfer of genes could greatly simplify traditional breeding procedures and overcome some of the inherent genetic problems, which otherwise would not be achievable through conventional methods. Indeed, more recently, orchids have been the subject of new areas of research, including functional genomics, proteomics, and metabolomics. The successful application of these new approaches to improve traits requires a reliable and reproducible transformation technique. The development and remarkable achievements of biotechnology in orchids during the past decade are reviewed, as are potential areas of research for the improvement of orchids.

*Corresponding author.

[†]*Agricultural Biotechnology Research Center, Academia Sinica, Taipei, Taiwan.*

13.1 Introduction

Orchids are the largest group of flowering plants and comprise approximately 750 genera, more than 25,000 identified species, and 120,000 hybrids. They represent the upper class in floriculture as cut flowers and potted plants worldwide.[1,2] According to the 2005 floriculture crops survey of the United States Department of Agriculture, orchid sales increased by approximately 12% over the previous year and currently account for the second highest sales in potted flowering plants (behind poinsettias), with a wholesale value of nearly US$144 million. The demand for orchids is increasing tremendously; in 2005, more than 18 million orchids were sold in the United States alone (http://www.orchidweb.org).

In response to ever-increasing consumer demand, the creation of new varieties with unique flower patterns, colors, shapes, fragrance, and resistance against pests and diseases is highly desirable.[1] Recently, Taiwan has emerged as a leader in the breeding and export of orchids. According to statistics from the Bureau of Foreign Trade in Taiwan, sales in 2003 accounted for approximately US$20 million. As a token of encouragement, the United States gave Taiwan permission to export *Phalaenopsis* grown in approved media to US soil in 2004. Currently, approximately 200 hectares of land in Taiwan are devoted to cultivating orchids, with most orchid farms being located in Yunlin, Chiayi, and Tainan counties of southern Taiwan.[2] To boost the floriculture industry in Taiwan, the government has invested nearly US$25 million in infrastructure development across the country.[3]

Traditionally, classical breeding has been used to introduce new traits for creating new hybrids in many orchid species. However, the technique is tedious and selected offspring are maintained throughout vegetative propagation to ensure genetic identity. Genetic improvement in orchids by sexual hybridization is hampered because of long juvenile periods and reproductive cycles. In addition, the available gene pool for new traits in orchids is limited because of the genetic background of parents. Nevertheless, rapid and dynamic progress in molecular biology and biotechnology has allowed for producing new orchid varieties with novel characteristics, such as flower pigmentation as well as disease and pest resistance. Sustaining both supply and demand of orchids in the future will depend more heavily on the development and deployment of new technologies, including genetic engineering.[2,4,5]

Gene transformation is an essential tool, both for the experimental investigation of gene function and the improvement of orchids, either by enhancing existing traits or introducing new genes.[5] Introducing and expressing foreign genes stably into the orchid genome by different means (bacteria/viruses), with *Agrobacterium*-mediation or microprojectile bombardment, is now possible, and many aspects of plant physiology and biochemistry that cannot be addressed easily by any other experimental means can be investigated by the analysis of gene function and regulation in transgenic orchids.

13.2 Foreign DNA Transfer Methods

In the last century, traditional plant breeding techniques have been used to increase many agronomic traits and the commercial value of orchids. The main drawback of traditional plant breeding is that it relies on the use of germplasm of the same or closely related species, which is sometimes a serious limiting factor.[5] In addition, progress is time-consuming and relies on the extensive use of natural resources. Gene transformation is an important tool in biotechnology and is based on the introduction of DNA into totipotent plant cells, followed by the regeneration of such cells into whole fertile plants. Early developments arising from transgenic techniques in plants in the mid-1970s produced tomatoes, corn, soybean, and canola with modified traits. Since the 1980s, these techniques have matured, become more target-specific, and generated transgenic plants in most major agricultural crops.

Particularly, the last two decades have seen significant developments in plant transformation technologies for the insertion of foreign DNA into plant genomes.[6] Among available techniques for gene transformation, the most popular systems are *Agrobacterium tumefaciens*-mediated transformation and microparticle bombardment (biolistics), with which transgenic plants have been successfully established in a variety of plant species such as maize, rice, wheat, and barley. Compared with the growing number of reports of genetically engineered crops in the literature, very few studies have demonstrated engineered plants for orchids. The optimization of efficient DNA delivery, successful regeneration, and suitable selection system are, therefore, prerequisites for most orchid transformation systems.

The three major plant transformation systems routinely employed in orchid transformation — *Agrobacterium tumefaciens*-mediated transformation, microprojectile bombardment, and direct gene transfer into protoplasts — have been used in the transfer of orchids, but only in a few species compared with agronomic plants.[7–11] So, research into improving existing or establishing new transformation techniques for orchids is essential.

13.3 Prerequisites for DNA Transfer into Orchids

13.3.1 *Explant type and regeneration capacity*

A reliable gene transformation system depends on the regeneration capacity of putatively transformed tissues so that various potential genes and their functions can be explored. In most orchids, protocorm-like bodies (PLBs) serve as target tissue for gene transformation: as the origin of these PLBs are single somatic cells, they are easy to root, presumed to be genetically uniform, and can be induced efficiently from various somatic tissues including young leaves, stem segments, or flower stalks, regardless of cultivar. Routine gene transformation systems have been developed in *Oncidium*, *Dendrobium*, and *Phalaenopsis*, with PLBs used as starting material.[10,12–15] However, Belarmino and Mii[7] used cell clumps/aggregates in *Phalaenopsis* transformation. Yu *et al.*[13] regenerated *Dendrobium* transgenic plants by using thin-section explants from PLBs, whereas calli or PLBs were used as target explants in *Dendrobium phalaenopsis* and *D. nobile* transformation.[10]

Several media compositions have been standardized for orchid cultivars for successful regeneration of PLBs and subsequent regeneration of putatively transformed plants. However, the nutritional requirement in orchid transformation depends on the species or variety. For example, in *Oncidium* transformation, a G10 medium supplemented with 4.3 g/L MS salts, 1 g/L tryptone, 20 g/L sucrose, 1 g/L charcoal, 65 g/L potato tubers, and 3 g/L phytagel (pH 5.4) is routinely used.[12] T2 medium (3.5 g Hyponex No. 1 [N:P:K 6:6.5:19; Hyponex Co. Ltd., Japan], 1 g tryptone, 0.1 g citric acid, 20 g sucrose, 1 g charcoal, 20 g sweet potato, 25 g banana and 3 g phytagel, pH 5.5) was successful in *Phalaenopsis* transformation.[1,14,15] Growth and morphogenesis *in vitro* are said to be controlled by the interaction and balance between growth regulators in the medium and growth substances produced endogenously by cultured cells.[16] The use of auxin and cytokinin concentration is unresolved in orchid transformation.

13.3.2 *Selection marker gene*

The proportion of totipotent cells that become transformed is low compared with untransformed cells. Most foreign genes introduced into the orchid genome do not confer a phenotype that can be conveniently used for selective propagation of transformed cells. For this reason, a selectable marker gene is introduced at the same time as the nonselectable foreign DNA to enable the survival of transformed cells in the presence of a particular chemical, the selective agent, which is toxic to nontransformed cells.

Currently, more than 50 selection systems have been reported, but only a few are frequently used in plant transformation.[17,18] Examples are those that used a gene encoding neomycin phosphotransferase II (*NPT* II), hygromycin phosphotransferase II (*HPT* II), and phosphinothricin acetyltransferase (*bar*) to develop the first generation of transgenic crops, regardless of tissue systems,[17,19] in order to confer antibiotic resistance with the gene of interest. The most commonly used antibiotics are hygromycin and kanamycin. Resistance to these antibiotics is conferred by insertion and expression of a gene encoding hygromycin phosphotransferase II (*HPT II*) under the control of Cauliflower mosaic virus (CaMV) 35S promoter. In direct DNA transfer methods, the selectable marker and nonselected transgene(s) may be linked on the same cointegrate vector or may be introduced on separate vectors (cotransformation). Both strategies are suitable because exogenous DNA, whether homogeneous or a mixture of different plasmids, predominantly integrates at a single locus.

Experiments in our laboratory involving the pCAMBIA (Center for the Application of Molecular Biology of International Agriculture, Black Mountain, Australia) binary vector series showed the most consistent results in *Oncidium* and *Phalaenopsis* transformation.[8,14,15] Screenable markers or reporter genes are widely used, particularly when transformation procedures are being optimized. The most commonly used gene in higher plants is *gusA*, which encodes β-glucuronidase (GUS) (Tables 13.1 and 13.2). Expression and activity of this enzyme are easily

Table 13.1. *Agrobacterium*-Mediated Transformation in Orchids

Species/Variety	Target Tissue	*Agrobacterium* Strain	Selection Marker Gene	Reporter/ Foreign Gene	Transformation Efficiency and Remarks	Reference No.
Dendrobium (UH800, UH44 and Thai hybrid M61)	PLBs	LBA4301	*vir*		Presence of coniferyl alcohol in PLBs acts as *vir* gene inducer	22
Phalaenopsis (True Lady A76-13, Brother Mirage A79-69, Asian Elegance B79-11 and Taisuco Kaaladian F80-13)	Transversely bisected PLBs	EHA105 (pMT1)	*NPT* II	Intron-GUS	100% GUS gene expression in Asian Elegance B79-11 and Taisuco Kaaladian F80-13, 50–80% in True Lady A76-13 and lowest in Brother Mirage A79-69	23
Phalaenopsis (*Doritaenopsis* Coral Fantasy X *Phalaenopsis*) (Baby Hat X Ann Jessica)	Cell clumps	LBA4404 (pTOK233) and EHA101 (pIG121Hm)	*NPT* II and *HPT* II	Intron-Gus	More than 100 hygromycin-resistant plantlets produced and transformation was confirmed by histochemical GUS assay, PCR and Southern hybridization analysis	7
Dendrobium (*Dendrobium* Somasak X *Dendrobium* Suzie Wong)	PLBs	LBA4404	*NPT* II	Orchid *DOH*1 antisense gene	Expression of *DOH*1 antisense gene caused abnormal shoot development and presence of transgene in transgenic plants confirmed by genomic PCR and Southern hybridization	8

(*Continued*)

Table 13.1. (*Continued*)

Species/Variety	Target Tissue	*Agrobacterium* Strain	Selection Marker Gene	Reporter/ Foreign Gene	Transformation Efficiency and Remarks	Reference No.
Phalaenopsis (T0, T5, T10 and Hikaru)	Intact and transversely bisected PLBs	LBA4404	*NPT* II	Intron-GUS	Transgenic plants derived from T0 and Hikaru showed 70% GUS expression and 10% of transformation efficiency was observed in Hikaru PLBs	24
Dendrobium nobile	PLBs	AGL1 and EHA105	*HPT* II	Intron-GUS	18% transformation efficiency obtained with 73 stably transformed lines and transgenes confirmed by Southern blot and GUS histochemical assay	9
Oncidium (Sherry Baby cultivar OM8)	PLBs	EHA105 and LBA4404	*HPT* II	Intron-GUS	Among 1000 inoculated PLBs, 108 putatively transformed PLBs were proliferated on hygromycin selection medium (10%). A total of 28 independent transgenic orchid plants were obtained, from which six transgenic lines were confirmed by Southern, Northern, and Western blot analyses	12

(*Continued*)

Table 13.1. (*Continued*)

Species/Variety	Target Tissue	*Agrobacterium* Strain	Selection Marker Gene	Reporter/ Foreign Gene	Transformation Efficiency and Remarks	Reference No.
Oncidium (Sherry Baby cultivar OM8)	PLBs	EHA105	*HPT* II	*pflp*, GFP and Intron GUS	Demonstrated *pflp* as selection marker gene and *Erwinia carotovora* as selection agent. Out of 32 independent transgenic lines, 9 were randomly selected and confirmed by Southern and Northern blot analyses	25
Oncidium (Sherry Baby cultivar OM8)	PLBs	EHA105	*HPT* II	*pflp*, GFP and Intron GUS	Among 17 independent transgenic orchid lines, 6 (GUS) positive were randomly selected and confirmed by Southern, Northern, and Western blot analyses. Transgenic plants showed enhanced resistance to *E. carotovora*, even when the entire plant was challenged with the pathogen	26

(*Continued*)

Table 13.1. (*Continued*)

Species/Variety	Target Tissue	*Agrobacterium* Strain	Selection Marker Gene	Reporter/ Foreign Gene	Transformation Efficiency and Remarks	Reference No.
Phalaenopsis (TS97K)	PLBs transformed with CymMV cDNA (gene stacking)	EHA105	HPT II	*pflp*, GFP and Intron GUS	Transgene integration confirmed by Southern and Northern blot analysis for both *CP* and *pflp* genes. Transgenic lines exhibited enhanced dual disease resistance to CymMV and *E. carotovora*	15
Phalaenopsis (S122-2 × S153 and S153 × S119-4)	Protocorms from germinating seeds	EHA101 (pIG121Hm)	*HPT* II	Intron-GUS	A total of 88 transgenic plants, each derived from an independent protocorm, was obtained from ca.12,500 mature seeds 6 months after infection with *Agrobacterium* and integration of *HPT* II gene confirmed by Southern blot analysis	11
Phalaenopsis (Wataboushi '#6.13)	Embryonic cell suspension culture	EHA101 (pEKH-WT)	*NPT* II and *HPT* II	Rice wasabi defensine	Integration and expression of defensin gene was confirmed by PCR, Southern, and Western blot analyses. Most transgenic plants showed strong resistance to *E. carotovora*	27

Table 13.2. Particle Bombardment/Direct Gene Transformation in Orchids

Species/Variety	Target Tissue	Selection Marker Gene	Reporter/ Foreign Gene	Transformation Efficiency and Remarks	Reference No.
Dendrobium (*Dendrobium* × Jaquelyn Thomas hybrids)	Protocoms	*NPT* II	Papaya ringspot virus (PRV) coat protein (CP) gene	Obtained 13 kanamycin resistant plants and confirmed foreign gene (*NPT* II) by genomic PCR and Southern blot analyses	37
Dendrobium (White angel)	Calli cultured in liquid medium	*NPT* II	Luciferase	Transformed tissues expressing luciferase were detected, allowed to regenerated into complete plant on selection medium and integration of transgene confirmed by Southern and Northern blot analyses	29
Phalaenopsis (Danse × Happy Valentine)	PLBs	*NPT* II and *bar*	Intron-GUS	About 7 bialophos-resistant plantlets were obtained from 622 bombarded PLBs and only one showed GUS expression	39
Dendrobium (hybrid 'MiHua')	Protocorms	*HPT* II	Intron-GUS	15 hygromycin-resistant lines recovered on selection medium and integration of foreign genes was confirmed by GUS histochemical assay and Southern blot analysis	13
Cymbidium	PLBs	*NPT* II	Intreon-GUS	About 85% of GUS positive shoots were obtained and further transgenic nature confirmed by genomic PCR and Southern blot analyses of the PCR product	40

(*Continued*)

Table 13.2. (*Continued*)

Species/Variety	Target Tissue	Selection Marker Gene	Reporter/ Foreign Gene	Transformation Efficiency and Remarks	Reference No.
Brassia, Cattleya and Doritaenopsis	PLBs	*bar*		Selection of putative transformants was accomplished by using bialaphos. The presence of bar gene was confirmed by PCR, Southern, and Northern blot analyses	20
Dendrobium phalaenopsis Banyan Pink and *D. nobile*	Calli	*HPT* II	Intron-GUS	About 12% (*D. Phalaenopsis*) and 2% (*D. nobile*) transformation efficiency was achieved and integration of foreign gene confirmed by Southern and Northern blot analyses	10
Phalaenopsis TS444 [(New Eagle × Pinlong Cinderella) × Dtps. Taisuco Red]	Petels	—	Flavonoid-3′,5′-hydroxylase gene from *Phalaenopsis*	Transient transformation achieved by particle bombardment and the transgenic petals changed from pink to magenta, demonstrated the role of flavonoid-3′,5′-hydroxylase gene in anthocyanin pigment synthesis	43
Phalaenopsis "TS340" (*P. Taisuco* Kochdiam × *P. Taisuco* Kaaladian)	PLBs	*HPT* II	CymMV coat protein (*CP*) cDNA	Among 13 transgenic plants confirmed by Southern blot analysis, most transgenic lines showed CymMV protection. Nuclear run-on and small interfering RNA (siRNAs) analyses showed that CymMV resistance was RNA-mediated through a posttranscriptional gene silencing mechanism (PTGS) in the silenced transgenic orchid plants.	14

(*Continued*)

Table 13.2. (*Continued*)

Species/Variety	Target Tissue	Selection Marker Gene	Reporter/ Foreign Gene	Transformation Efficiency and Remarks	Reference No.
Dendrobium (Hickam Deb)	Protocorms	*HPT* II	CymMV coat protein (*CP*) cDNA	Presence of the transgene confirmed by PCR, Southern, Northern, and Western blot analyses. Transgenic plants harboring the CymMV CP gene expressed a very low level of virus accumulation 4 months postinoculation with CymMV	42
Phalaenopsis (TS97K)	PLBs	*HPT* II	CymMV coat protein (*CP*) cDNA	PLBs transformed with CymMV coat protein cDNA (CP) were then retransformed with sweet pepper ferredoxin-like protein cDNA (*pflp*) by *Agrobacterium tumefaciens*, to enable expression of dual (viral and bacterial) disease resistant traits. Double transformants confirmed by molecular analysis and exhibited enhanced dual disease resistance	14
Oncidium (Sharry Baby "OM8")	PLBs (pretreated with sucrose)	*HPT* II	*pflp*	Sucrose pretreatment enhanced the regeneration of PLBs, single-cell embryogenesis, and transformation efficiency	41

detectable by use of both histochemical and fluorigenic substrates. This reporter gene was successfully used in the transformation of different orchids.[10,11] In some cases, the herbicide resistance (*bar*) gene has also been useful in the selection of transgenic orchids.[20]

13.4 General Protocol for *Agrobacterium*-Mediated Transformation in Orchids

The landmark discovery of *A. tumefaciens*[21] provided a natural gene transfer mechanism to introduce and express DNA stably in different plant species, regardless of their origin. Several hybrid orchids, which are difficult for *A. tumefaciens* infection, have been transformed by direct gene gun or microparticle bombardment. However, gene transfer techniques have been established only in a few orchids.[7,10,11,13] In general, *Agrobacterium*-mediated transformation in orchids was achieved by cocultivating a virulent *A. tumefaciens* strain containing a recombinant Ti plasmid with PLBs/cell clumps, from which complete plants were regenerated.

Although *Agrobacterium*-mediated transformation needs to be optimized for different species, the *Agrobacterium* infection can be promoted under conditions that induce virulence, such as use of acetosyringone (AS) or a-hydroxyacetosyringone, acidic pH, and appropriate incubation temperature. After coculture for 2 to 3 days, PLBs are transferred to a medium containing selective agents to eliminate non-transformed plant cells, antibiotics to kill the *Agrobacterium*, and hormones to induce shoot and root growth. After a few months, shoots develop from transformed PLBs. These can be removed and transferred to a rooting medium, and then shifted to soil. Most current protocols for the *Agrobacterium*-mediated transformation of orchids involve variations in the starting material. In *Phalaenopsis*, using PLBs as target tissue, we have developed a reliable and reproducible transformation protocol (Fig. 13.1).

13.4.1 *Recent advances in Agrobacterium-mediated transformation*

Several plant phenolic compounds, including AS, coumaryl alcohol, and sinapyl alcohol, are known as *vir* gene inducers; but in orchids, their

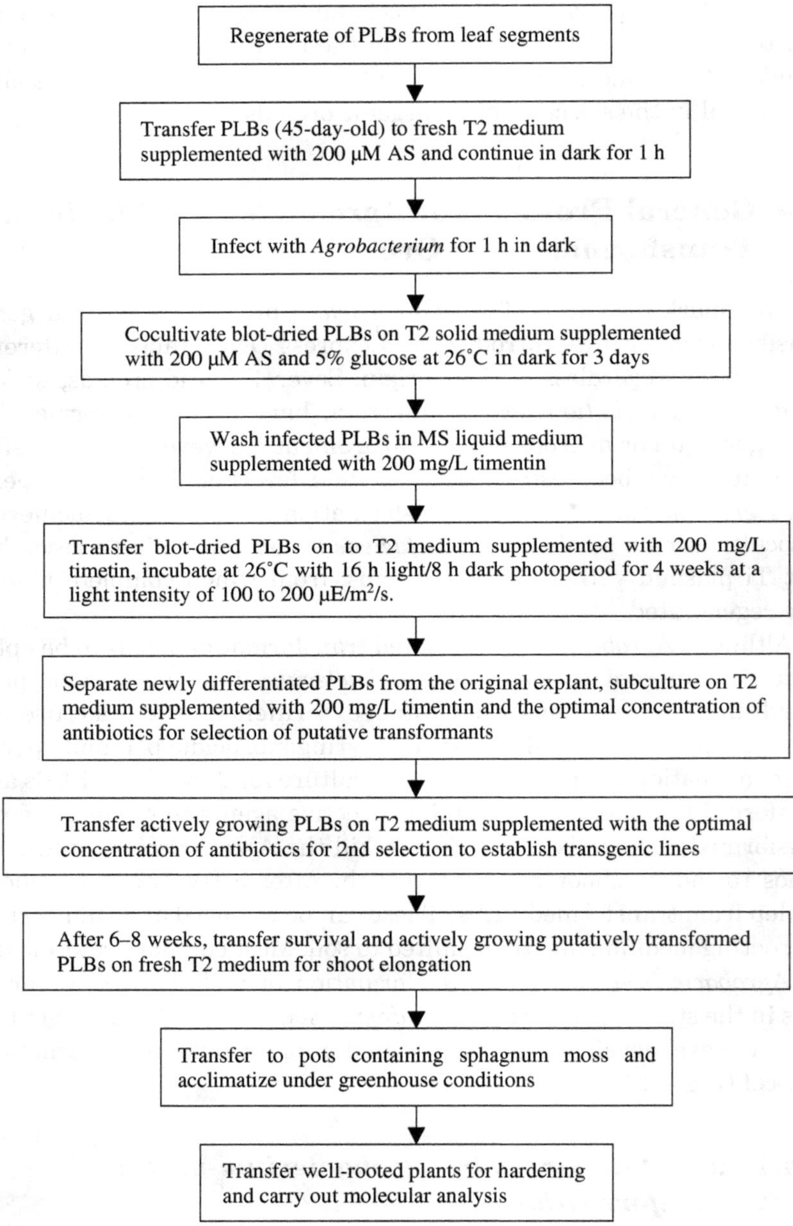
Regenerate of PLBs from leaf segments
Transfer PLBs (45-day-old) to fresh T2 medium supplemented with 200 μM AS and continue in dark for 1 h
Infect with *Agrobacterium* for 1 h in dark
Cocultivate blot-dried PLBs on T2 solid medium supplemented with 200 μM AS and 5% glucose at 26°C in dark for 3 days
Wash infected PLBs in MS liquid medium supplemented with 200 mg/L timentin
Transfer blot-dried PLBs on to T2 medium supplemented with 200 mg/L timetin, incubate at 26°C with 16 h light/8 h dark photoperiod for 4 weeks at a light intensity of 100 to 200 μE/m²/s.
Separate newly differentiated PLBs from the original explant, subculture on T2 medium supplemented with 200 mg/L timentin and the optimal concentration of antibiotics for selection of putative transformants
Transfer actively growing PLBs on T2 medium supplemented with the optimal concentration of antibiotics for 2nd selection to establish transgenic lines
After 6–8 weeks, transfer survival and actively growing putatively transformed PLBs on fresh T2 medium for shoot elongation
Transfer to pots containing sphagnum moss and acclimatize under greenhouse conditions
Transfer well-rooted plants for hardening and carry out molecular analysis

Fig. 13.1. Different stages of *A. tumefaciens*-based transformation in *Phalaenopsis*.

presence and characteristics were not well understood until Nan *et al.*[22] identified these compounds in high amounts in light-grown *Dendrobium* PLBs. The authors demonstrated activation of the *Agrobacterium vir* gene in the presence of these phenolic compounds, and revealed the possibility of *Agrobacterium*-mediated transformation in orchids. These observations come as a great boon to orchid biologists to initiate gene transformation studies in various commercial varieties.

The feasibility of using *Agrobacterium*-mediated transformation system in orchid transformation was not explored until Hsieh *et al.*[23] first reported the transformation of different varieties of *Phalaenopsis* with *NPT* II and GUS genes using PLBs as target tissue (Table 13.1). Subsequently, other authors reported success with *Agrobacterium*-mediated transformation in orchids. Belarmino and Mii[7] reported *Phalaenopsis* transformation with a GUS gene and confirmed the transgenic nature of developed plants by histochemical GUS assay, PCR analysis, and Southern hybridization. Use of this transformation system successfully produced more than 100 hygromycin-resistant plantlets within 7 months following infection of cell aggregates. *Dendrobium* was transformed with *DOH1* by use of thin-section explants from PLBs.[13]

Chai *et al.*[24] evaluated two types of explants, intact and transversely bisected PLBs, in *Phalaenopsis* transformation, and observed that transversely bisected PLBs responded better than intact ones in terms of efficiency of multiplication and transformation. A high efficiency of transformation (18%) in *Dendrobium* was reported, with 73 stably transformed lines accomplished in the presence of 100 μM AS and hygromycin as the selection marker.[10] Liau *et al.*[12] first reported a transformation method suitable for improving *Oncidium* by using PLBs derived from protocorms as target tissue. Among 1000 inoculated PLBs, 108 putatively transformed PLBs were proliferated on hygromycin selection medium. A total of 28 independent transgenic orchid plants were obtained, from which 6 transgenic GUS-positive lines were randomly selected and confirmed by Southern, Northern, and Western blot analyses. This successfully demonstrated the integration and expression of foreign DNA in the *Oncidium* genome. Soonafter, the same group transformed sweet pepper ferredoxin-like protein (*pflp*) cDNA, with a *HPT* II and *GUS* coding sequence, into PLBs of *Oncidium*, and demonstrated the applicability of the *pflp* gene as an antibiotic-free selection marker in *Oncidium*[25]; intriguingly, transgenic plants showed enhanced resistance to *Erwinia carotovora*, which causes soft rot disease, even when the entire plant was challenged with the pathogen.[26]

Recently, Chan *et al.*[15] used "gene stacking" in *Phalaenopsis* by double transformation events. *Phalaenopsis* PLBs originally transformed with CymMV coat protein cDNA (*CP*) were then transformed with *pflp* cDNA by *A. tumefaciens* infection to enable the expression of dual (viral and bacterial) disease-resistant traits. The immature protocorms developed from *Phalaenopsis* seeds were also demonstrated to be a potential starting source for *Agrobacterium*-mediated transformation.[11] However, this protocol may not have broader application because most of the commercial hybrids are sterile in nature. Interestingly, 5 day-old embryogenic suspension cells of *Phalaenopsis* were used as starting material in rice wasabi defensin gene transformation, and the authors claimed that most of the transgenic plants exhibited enhanced resistance to *E. carotovora*.[27]

13.5 General Conditions for *A. Tumefaciens*-Mediated Transformation

13.5.1 *Effect of feeder cells/acetosyringone on transformation efficiency*

Acetosyringone (AS), a phenolic compound produced in metabolically active plant cells or released by wounded cells, induces the expression of the *vir* genes on the Ti plasmid of *A. tumefaciens*, which in turn facilitates T-DNA transfer into plant cells.[28] Although orchid cells are capable of producing certain phenolic compounds at a low level, such production is not sufficient for high and consistent transformation events.[22]

The feasibility of AS in orchid transformation was demonstrated by Hsieh *et al.*[23] in *Phalaenopsis* transformation; the authors added 100 μM AS and regeneration medium to the *Agrobacterium* culture prior to infection to enhance transformation events. In concordance with the above observations, Chia *et al.*[29] and Mishiba *et al.*[11] used 100 μM AS during the infection and cocultivation stages in *Phalaenopsis* transformation. In *Dendrobium*, Men *et al.*[10] observed higher GUS gene expression when PLBs were cocultivated with 100 μM AS-activated *A. tumefaciens* for 2 to 3 days. Interestingly, Chan *et al.*[15] reported the use of 200 μM AS and 5% glucose during cocultivation and found enhanced transformation efficiency. Belarmino and Mii[7] infected explants with 200 μM AS-treated *Agrobacterium* and included 500 μM AS in cocultivation medium. In *Oncidium* transformation, the use of 1 mL of tightly packed 3-day-old freshly subcultured tobacco suspension cells on G10 medium supplemented with 200 μM AS as a cocultivation medium and dark incubation for 3 days at 26°C significantly triggered the *A. tumefaciens vir* genes, which in turn resulted in high transformation efficiency. Furthermore, these results were consistent and reproducible.[12] However, in the absence of AS, only a few transgenic plants were obtained. The use of feeder cells from species such as tobacco petunia, and tomato as a phenolic source to enhance the transformation rate still needs to be optimized for various commercial orchids.

13.5.2 *Agrobacterium strain and density*

The choice of *Agrobacterium* strain greatly influences the transformation success in orchids. A wide range of *A. tumefaciens* strains used include LBA4301,[22] EHA105 (pMT1),[23] LBA4404 (pTOK233),[8,24] EHA101 (pIG121Hm),[11] and EHA101 (pEKH-WT).[27] However, others have advocated that the success of transformation efficiency is not influenced by the *Agrobacterium* strain alone.[7,30] In *Dendrobium*, Men *et al.*[10] examined two *Agrobacterium* strains, AGL1 and EHA105, and observed a high level of GUS staining in AGL1-infected PLBs. Simultaneously, Liau *et al.*[26] using EHA105 and LBA4404, found no difference in transformation efficiency. However, the shoot formation rate of EHA105-transformed PLBs was higher than that of LBA4404-treated PLBs. In addition, after hygromycin selection, the antibiotics used (timentin/cefatoxime) did not always control the overgrowth of LBA4404 and resulted in a low rate of shoot formation.

Therefore, in general, the tissue culture conditions and choice of bacteriostatic substances employed to suppress the overgrowth of *Agrobacterium* are important factors in orchid transformation. Although the bacterial density was not thoroughly examined in most of the transformation studies, most authors used diluted *Agrobacterium* at an A_{600} of 0.2–1.0 for improved transformation efficiency. Thus, *Agrobacterium* density is also a key factor for successful transformation.[11,23–26]

13.5.3 *Effect of selection agents*

In general, *Agrobacterium*-mediated transformation in orchids is preceded by infection, 2 to 3 days of cocultivation, and selection of putative transformants on selection medium supplemented with antibiotics and bacteriostatic substances to cull untransformed tissues and *A. tumefaciens*. The ideal antibiotic for inhibiting *Agrobacterium* growth should be highly effective, inexpensive, not have a negative effect on plant growth and regeneration, and be stable in culture.[31] The selective effects of different antibiotics in plant transformation have been reported for several species.[32] In orchid transformation, most authors used different concentrations of cefotaxime/carbenicillin as antibiotics for suppressing *Agrobacterium* growth after cocultivation.[7–9,23,24,27] Mishiba *et al.*[11] used meropenem (β-lactam antibiotics) to remove excess *Agrobacterium* from PLBs.

In our lab, we standardized a two-step transformation procedure in *Oncidium*: after infection, PLBs were cocultivated for 3 days; then, infected PLBs were transferred to a medium supplemented with low concentrations of timentin/cefatoxime to suppress *A. tumefaciens* growth and incubated at 26°C for a 16-h photoperiod at a light intensity of 100–200 $\mu Em^{-2}s^{-1}$ for a month. No selective antibiotic had been incorporated in the medium to select putatively transformed PLBs because the long-term transformation process might enhance transfer of foreign DNA from *A. tumefaciens* into actively growing PLBs. Subsequently, *A. tumefaciens* was completely avoided with high concentrations of timentin/cefatoxime and a selective agent was used to eliminate untransformed PLBs. We achieved approximately 10% transformation efficiency, with 108 antibiotic-resistant independent PLBs proliferated from 1000 infected PLBs.[12]

13.6 Microprojectile/Particle Bombardment or Direct Gene Transformation

A novel method for plant transformation was introduced by John Sanford and colleagues in 1987. The authors showed that projectile bombardment, with several modifications, was an efficient method for stable integration of foreign genes into the plant genome.[33] Subsequently, a gunpowder-based device was refined for a system based on high-pressure blasts of helium;[34] this is the only commercially available particle gun, which is marketed by Bio-Rad. Particle bombardment is a simple technique: a plasmid DNA is prepared by standard methods and precipitated onto tungsten or gold particles with use of $CaCl_2$, spermidine, and PEG to protect the DNA during precipitation; and then the particles are washed and suspended in ethanol before being fired against a retaining screen that allows the microprojectiles through to strike the target tissue.

Particle bombardment is widely used because it circumvents two major limitations of the *Agrobacterium* system. First, any hybrid orchid species and cell type can be transformed by this method because DNA delivery is controlled entirely by physical rather than biological parameters. The range of species transformable by particle bombardment is, therefore, restricted only by the competence of cells for regeneration. The technique is independent of genotype and, consequently, useful for the transformation of elite cultivars. However, careful optimization is required to tailor the method for different varieties or cell types and to achieve the highest efficiency transformation with the least cell damage. Important parameters include acceleration method, particle velocity (controlled by the discharge voltage and/or gas pressure), particle size, and use of different materials (tungsten, gold).[35] Second, particle bombardment allows the stable and heritable introduction of many different genes at once with the use of different plasmids, as these tend to concatemerize to form one DNA cluster that integrates at a single locus. Remarkably, the transgene silencing, instability, and rearrangements are more evident with particle bombardment than the *Agrobacterium* method of gene transformation.[36]

13.6.1 *Recent advances in particle bombardment in orchids*

Pioneering gene transformation experiments augmented with particle bombardment were successfully used to produce transgenic orchids with a marker/reporter gene in *Dendrobium*,[10,13,28,37] *Phalaenopsis*,[39] *Cymbidium*,[40] *Brassia*, *Cattleya*, and *Doritaenopsis*[20] (Table 13.2). Osmotic pretreatment of *Oncidium* PLBs on high sucrose medium before bombardment and transformation have enhanced single-cell embryogenesis and transformation efficiency.[41] This situation can also be achieved by partial drying or the addition of osmoticum (mannitol

Table 13.3. Other Direct Gene Transformation Methods in Orchids

Species/Variety	Target Tissue	Transformation Method	Reference No.
Dendrobium	Seeds	Imbibition in solution containing foreign DNA	44
Dendrobium	Ovary	Pollen-tube pathway	44
Dendrobium	Protocorms	Electro injection	44
Phalaenopsis	Ovary	Pollen-tube pathway	38

and/or sorbitol) to the culture medium. In general, stable transformation by any direct DNA transfer method occurs at a much lower frequency than transient transformation.

The direct gene transformation has also been successfully used in developing transgenic orchids resistant to Cymbidium mosaic virus in *Phalaenopsis*[14,15] and *Dendrobium*.[42] Interestingly, Chan *et al.*[15] demonstrated the usefulness of direct gene transformation in introducing multiple genes into the *Phalaenopsis* genome. Su and Hsu[43] reported transient transformation of *Phalaenopsis* petals with the putative cytochrome P450 gene. In addition, other direct gene transformation methods, including seed imbibition and pollen tube mediation, have been reported in *Dendrobium* and *Phalaenopsis*[38,44] (Table 13.3). However, these methods are not in frequent use because of low transformation efficiency.

13.7 Selection of Putative Transformants and Molecular Analysis

Selection of the transformed cells and subsequent regeneration of plants is the most important requirement in orchid transformation. The most commonly used antibiotics for selection of transformed orchid PLBs are hygromycin and kanamycin. Resistance to these antibiotics is conferred by insertion and expression of *HPT* II or *NPT* II under the control of a CaMV35S promoter.[7,12,13,23,24,37] Knapp *et al.*[20] demonstrated the usefulness of *bar* as a selection marker in *Brassia*, *Cattleya*, and *Doritaenopsis* transformation. Our laboratory experiments involving pCAMBIA binary vector series showed the most consistent results in *Phalaenopsis* and *Oncidium* transformation. The screenable markers or reporter genes are widely used, particularly when transformation procedures are being optimized. The most commonly used reporter gene in higher plants is GUS; the expression and activity of this enzyme are easily detectable by the use of both histochemical and fluorigenic substrates. This reporter gene was used successfully in the transformation of other orchids, including *Cymbidium*,[37] *Dendrobium*,[10] and *Phalaenopsis*.[11]

Furthermore, the available molecular tools such as PCR, Northern, Southern, and Western blot analyses have allowed for confirming transgene integration and expression and have clarified the outcome of such transformation experiments. Because orchids have a long flowering stage, the molecular analysis of T_1 progeny is time-consuming. Alternatively, T_0 transgenic lines can be thoroughly diagnosed for foreign gene integration by Southern, Northern, and Western blot analyses.[12] Southern and Northern blot analysis was used to confirm the transformation, copy number, and transcriptional activity of transgenes in putatively transformed orchids.

13.8 Expression of Genes with Potential Commercial Application

The main drawback of traditional plant breeding is that it relies on the use of germplasm of the same or closely related species, and the progress is time-consuming. The 21st century will likely pose a new set of challenges for orchid growers, as increased demand for cut flowers and potted plants will require quick production practices. Sustaining both supply and demand in the future will depend more heavily on the development and deployment of a range of new technologies, including biotechnology.

13.8.1 *Marker-free transgenic orchids*

To obtain maximum putatively transformed plants in each transformation event, effective selectable markers, antibiotics, or herbicide-resistant genes have been used to simplify detection procedures. The existing selection systems can be divided into two groups: conventional selection systems, the largest group, which rely on an antibiotic or a herbicide as the selective agent detoxified by a selective gene;[45] and positive selection systems, in which the selective agent is converted into a simple compound by the selective gene product and transformed cells have metabolic or developmental advantages.[46,47] In most of the orchid transformation protocols, the available methods for screening transgenic plants are based on antibiotics or herbicide-resistant marker genes such as *NPT* II, *HPT* II, and *bar* on media containing the selective agent.[7,8,20] However, the use of these selection marker genes is a matter of concern because of their possible toxicity or allergenicity to humans and other organisms. In addition, the presence of such resistance genes and their proteins is undesirable in transgenics. Furthermore, these selection systems are time-consuming, expensive, and labor-intensive, therefore, the development of alternative or new selection marker genes and improved selection procedures is essential for the orchid industry.

Sweet pepper ferredoxin-like protein (*pflp*), isolated from pepper, was successfully cloned and reported to have antimicrobial activity against *Pseudomonas syringae* by delaying the hypersensitive response through the release of a hairpin proteinaceous elicitor.[48] We developed a novel selection procedure for orchid transformation by introducing *pflp* as a selection marker into the *Oncidium* genome via *A. tumefaciens* infection and particle bombardment, and putatively transformed plants were screened by the use of the natural bacterial pathogen *E. carotovora* as a selection agent (Fig. 13.2). We were able to obtain approximately 32 independent transgenic lines without the use of an antibiotic selection agent.[25] Although the selection efficiency with *pflp* or hygromycin does not differ much, the time required for selecting transgenic plants by use of *pflp* is shorter. Therefore, to minimize the chances for somaclonal variation caused by long-term tissue culture selection procedures and to enhance the rooting ability of transgenic plants, always a serious problem in the antibiotic selection system, the use of *pflp* is preferable. Furthermore, the *pflp* system can be used to introduce economically important genes into any other orchid. Recently, Chan *et al.*[15] successfully demonstrated the practicality of the *pflp* system as a nonantibiotic selection system in *Phalaenopsis* transformation. Plants that are not natural hosts of *E. carotovora*, used as transformation materials, do not conform to the *pflp* selection system. However, the deployment of this system may substantially decrease concerns about the negative effect on

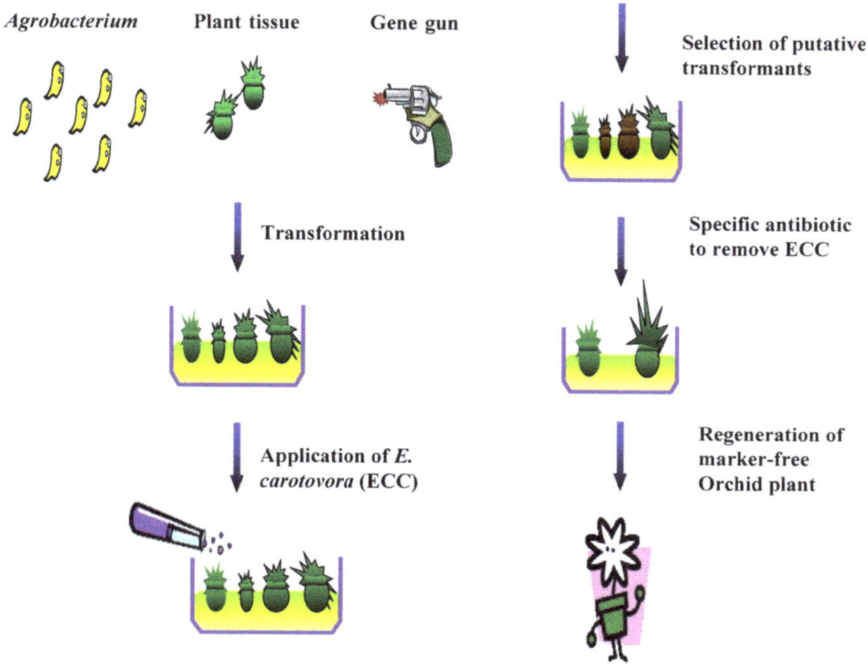

Fig. 13.2. Schematic representation of the development of marker-free transgenic orchids with *pflp* used as a marker gene and *Erwinia carotovora* (ECC) as a selection agent.

biodiversity with the use of antibiotics or herbicide-resistance marker genes and their presence in transgenic plants.

13.8.2 *Disease-resistant genes*

The efficient control of phytopathogens affecting orchid crops represents one of the major challenges in the floriculture industry worldwide. Several orchid species are highly susceptible to different pathogens, among them the devastating disease soft rot caused by *E. carotovora*, and this genus is considered to be the natural host. Advances in technology have allowed for introducing a novel gene(s) singly or in combination into the orchid genome for the dual purposes of transformation and disease resistance. For example, the overexpression of *pflp* in rice confers resistance against *Xanthomonas oryzae*.[49] We introduced an antimicrobial gene (*pflp*) into the *Oncidium* genome via *A. tumefaciens* infection. Application of bacteria on the leaves of transgenic and wild-type plants showed characteristic symptoms of water-soak regions followed by maceration, more evident in wild-type plants than in transgenic plants; resistance scoring at the whole plant level also revealed that *pflp* confers enhanced resistance to *E. carotovora*.[26]

Viral diseases in orchids, especially those caused by *Odontoglossum* ringspot tobamovirus (ORSV) and *Cymbidium* mosaic virus (CymMV), are a major concern among breeders because of their catastrophic effects on plant yield and market value. Liao *et al.*[14] successfully created transgenic *Phalaenopsis* orchid plants transformed with CymMV coat protein cDNA (*CP*), which conferred protection against CymMV infection. To enhance the resistance of orchids to both viral and bacterial phytopathogens, Chan *et al.*[15] used the "gene-stacking" phenomena in *Phalaenopsis* by double transformation events: *Phalaenopsis* PLBs, originally transformed with *CP*, were then transformed with *pflp* cDNA by the use of *A. tumefaciens* to enable the expression of dual (viral and bacterial) disease-resistant traits. Disease-resistant assays revealed transgenic plants with enhanced resistance to CymMV and *E. carotovora* infection. Recently, Chang *et al.*[39] created transgenic *Dendrobium* plants harboring the CymMV *CP* gene, which expressed a very low level of virus accumulation 4 months after inoculation with CymMV.

Large numbers of toxic compounds exist in higher plants to ensure various physical and biochemical defense barriers against pests and pathogens. Among these, defensin proteins form a well-defined group of low-molecular-weight cystein-rich polypeptides whose toxic effect on fungi and bacteria have been studied in model systems and are known to be expressed either by pathogen attack or elicitors. Recently, Sjahril *et al.*[27] successfully introduced the wasabi defensin gene into *Phalaenopsis* and recorded enhanced tolerance to *E. carotovora*. Similar strategies can be extended to other commercially important orchids for single- or multiple-disease resistance.

13.9 Conclusions and Future Prospects

As more genes are being isolated and understood in model systems and crop plants, more options are becoming available for the application of genetic modification in orchid improvement. However, such advances have not been fully realized in orchids, probably because most hybrid orchids are recalcitrant to tissue culture and transformation. So, research to increase transformation efficiency, develop new marker genes, introduce new genes alone or in combination by multigene transformation (gene stacking), and improve cryopreservation techniques would be extremely useful for orchid improvement.

Recent studies of the entire chloroplast genome sequencing (cpDNA) of *Phalaenopsis* are novel examples of the successful utilization of cutting-edge technology in orchids, and the information generated can be successfully utilized in chloroplast transformation in *Phalaenopsis* and other orchids.[50] Functional genomics studies of *Oncidium* and *Dendrobium* flower formation and initiation are promising and useful in terms of industrial as well as academic interests.[51–53] In addition, the information generated by Wu *et al.*[54] on *Dendrobium MYB* genes during flower development could be useful in exploring new genes involved in early/late flower development, subsequent functional analysis, and innovation of novel flower-specific promoters that will certainly benefit orchid improvement. Pathway engineering, involving the simultaneous introduction and concerted expression of multiple transgenes, will be featured prominently in future strategies for orchid transformation. Finally, the outputs from current genomics, proteomics, and metabolomics programs can be fully exploited with sophisticated methods to introduce large pieces of DNA into the orchid genome.

Acknowledgments

Updates are not intended to be comprehensive reviews, and the authors apologize to those colleagues whose works could not be cited because of the restricted number of references. The authors thank S.J. Yu, I.C. Pan, and C.H. Liau for graphics. This work was supported by grants from Academia Sinica as well as the National Science and Technology Program for Agricultural Biotechnology of the Republic of China.

References

1. Chan MT, Chan YL, Sanjaya. (2006) Orchids (Cymbidium spp., *Oncidium* and *Phalaenopsis*). In: Kang Wang (ed.), *Methods in Molecular Biology*, Vol. 344: Agrobacterium *Protocols*. 2/e, Vol. II, Humana Press Inc., Totowa, USA, pp. 331–337.

2. Chan MT, Sanjaya, Teixeira da Silva JA. (2006) *Oncidium* tissue culture, transgenics and biotechnology. In: Teixeira da Silva JA (ed.), *Floriculture, Ornamental and Plant Biotechnology: Advances and Topical Issues*, Vol. II, Global Science Books, Ltd., London, UK, pp. 193–198.
3. Chan MT, Sanjaya. (2006) Current trends in Taiwan's orchid biotechnology. *Global Plant Lett* **I:**2–3.
4. Teixeira da Silva AJ. (2004) Ornamental chrysanthemums: Improvement by biotechnology. *Plant Cell Tissue Organ Cult* **79:**1–18.
5. Tanaka Y, Katsumoto Y, Bruglier F, Mason J. (2005) Genetic engineering in floriculture. *Plant Cell Tissue Organ Cult* **80:**1–24.
6. Vain P. (2006) Global trends in plant transgenic science and technology (1973–2003). *Trends Biotechnol* **24:**206–211.
7. Belarmino MM, Mii M. (2000) *Agrobacterium*-mediated genetic transformation of a *Phalaenopsis* orchid. *Plant Cell Rep* **19:**435–442.
8. Yu H, Yang SH, Goh CJ. (2001) *Agrobacterium*-mediated transformation of a *Dendrobium* orchid with the class 1 knox gene DOH1. *Plant Cell Rep* **20:**301–305.
9. Men S, Ming X, Liu R, *et al.* (2003) *Agrobacterium*-mediated genetic transformation of a *Dendrobium* orchid. *Plant Cell Tissue Organ Cult* **75:**63–71.
10. Men S, Ming X, Wang Y, *et al.* (2003) Genetic transformation of two species of orchid by biolistic bombardment. *Plant Cell Rep* **21:**592–598.
11. Mishiba KI, Chin DP, Mii M. (2005) *Agrobacterium*-mediated transformation of *Phalaenopsis* by targeting protocorms at an early stage after germination. *Plant Cell Rep* **24:**297–303.
12. Liau CH, You SJ, Prasad V, *et al.* (2003) *Agrobacterium tumefaciens*-mediated transformation of an *Oncidium* orchid. *Plant Cell Rep* **21:**993–998.
13. Yu Z, Chen M, Nie L, *et al.* (1999) Recovery of transgenic orchid plants with hygromycin selection by particle bombardment to protocorms. *Plant Cell Tissue Organ Cult* **58:**87–92.
14. Liao LJ, Pan IC, Chan YI, *et al.* (2004) Transgene silencing in *Phalaenopsis* expressing the coat protein of *Cymbidium* mosaic virus is a manifestation of RNA-mediated resistance. *Mol Breed* **13:**229–242.
15. Chan YL, Lin KH, Sanjaya, *et al.* (2005) Gene stacking in *Phalaenopsis* orchid enhances dual tolerance to pathogen attack. *Transgenic Res* **14:**279–288.
16. George EF, Sherrington PD. (1984) *Plant Propagation by Tissue Culture — Handbook and Dictionary of Commercial Laboratories*. Exgetics Ltd., England.
17. Miki B, McHugh S. (2004) Selectable marker genes in transgenic plants: Applications, alternatives and biosafety. *J Biotechnol* **107:**193–232.

18. Tian L. (2006) Markers gene removal from transgenic plants. In: Teixeira da Silva JA (ed.), *Ornamental and Plant Biotechnology*, Vol. II, Global Science Books, London, pp. 26–29.
19. Haldrup A, Petersen S, Okkels F. (1998) Plant xylose isomerase gene from *Thermoanaerobacterium thermosulfurogenes* allows effective selection of transgenic plant cells using D-xylose as the selection agent. *Plant Mol Biol* **37:**287–296.
20. Knapp JE, Kausch AP, Chandlee JM. (2000) Transformation of three genera of orchid using the bar gene as a selectable marker. *Plant Cell Rep* **19:**893–898.
21. Chilton MD, Drummond MH, Merlo DJ, *et al.* (1977) Stable incorporation of plasmid DNA into higher plant cells: The molecular basis of crown gall tumorigenesis. *Cell* **11:**263–271.
22. Nan GL, Tang CS, Kuehnle AR, Kado CI. (1997) *Dendrobium* orchids contain an inducer of *Agrobacterium* virulence genes. *Physiol Mol Plant Pathol* **51:**391–399.
23. Hsieh RM, Chen WH, Lin YS, *et al.* (1997) *Agrobacterium tumefaciens*-mediated transformation of *Phalaenopsis* orchid. *Rept Taiwan Sugar Res Inst* **155:**41–54.
24. Chai ML, Xu CJ, Senthil KK, *et al.* (2002) Stable transformation of protocorm-like bodies in *Phalaenopsis* orchid mediated by *Agrobacterium tumefaciens*. *Sci Hortic* **96:**213–224.
25. You SJ, Liau CH, Hauang HE, *et al.* (2003) Sweet pepper ferredoxin-like protein (*pflp*) gene as a novel selection marker for orchid transformation. *Planta* **217:**60–65.
26. Liau CH, Lu JC, Prasad V, *et al.* (2003) The sweet pepper ferredoxin-like protein (*pflp*) conferred resistance against soft rot disease in *Oncidium* orchid. *Transgenic Res* **12:**329–336.
27. Sjahril R, Chin DP, Khan RS, *et al.* (2006) Transgenic *Phalaenopsis* plants with resistance to *Erwinia carotovora* produced by introducing wasabi defensin gene using *Agrobacterium* method. *Plant Biotechnol* **23:**191–194.
28. Bolton GW, Nestor EW, Gordon MP. (1986) Plant phenolic compounds induce expression of the *Agrobacterium tumefaciens* loci needed for virulence. *Science* **232:**983–985.
29. Chia TF, Chan YS, Chua NH. (1994) The firefly luciferase gene as a non-invasive reporter for *Dendrobium* transformation. *Plant J* **6:**441–446.
30. Hiei Y, Komari T, Kuba T. (1997) Transformation of rice mediated by *Agrobacterium tumefaciens*. *Plant Mol Biol* **35:**205–218.
31. Cheng ZM, Schnurr JA, Kapaun JA. (1998). Timentin as an alternative antibiotic for suppression of *Agrobacterium tumefaciens* in genetic transformation. *Plant Cell Rep* **17:**646–649.

32. Hammerschlag FA, Zimmerman RH, Yadava *et al.* (1997) Effect of antibiotics and exposure to an acidified medium on the elimination of *Agrobacterium tumefaciens* from apple leaf explants and on shoot regeneration. *J Am Hortic Soc* **112:**758–763.
33. Klein TM, Fromm M, Weissinger A, *et al.* (1988) Transfer of foreign genes into intact maize cells with high velocity microprojectiles. *Proc Natl Acad Sci USA* **85:**4305–4309.
34. Ye G-N, Daniell H, Stanford JC. (1990) Optimization of delivery of foreign DNA into higher plant chloroplasts. *Plant Mol Biol* **15:**809–819.
35. Christou P. (1994) Particle gun mediated transformation. *Curr Opin Biotechnol* **4:**135–141.
36. Birch RG. (1997) Plant transformation: Problems and strategies for practical application. *Annu Rev Plant Physiol Plant Mol Biol* **48:**297–326.
37. Kuehnle AR, Sugii N. (1992) Transformation of *Dendrobium* orchid using particle bombardment of protocorms. *Plant Cell Rep* **11:**484–488.
38. Hsieh YH, Huang PL. (1995) Studies on genetic transformation of *Phalaenopsis* via pollen tube pathway. *J Chin Soc Hortic Sci* **41:**309–324.
39. Anzai H, Ishii Y, Shichinohe M, *et al.* (1996) Transformation of *Phalaenopsis* by particle bombardment. *Plant Tissue Cult Lett* **3:**265–271.
40. Yang J, Lee H, Shin DH, *et al.* (1999) Genetic transformation of *Cymbidium* orchid by particle bombardment. *Plant Cell Rep* **18:**978–984.
41. Li SH, Kuoh CS, Chen YH, *et al.* (2005) Osmotic sucrose enhancement of single-cell embryogenesis and transformation efficiency in *Oncidium*. *Plant Cell Tissue Organ Cult* **81:**183–192.
42. Chang C, Chen YC, Hsu YH, *et al.* (2005) Transgenic resistance to *Cymbidium* mosaic virus in *Dendrobium* expressing the viral capsid protein gene. *Transgenic Res* **14:**41–46.
43. Su V, Hsu BD. (2003) Cloning and expression of a putative cytochrome P450 gene that influences the colour of *Phalaenopsis* flowers. *Biotechnol Lett* **25:**1933–1939.
44. Nan GL, Kuehnle AR. (1995) Genetic transformation in *Dendrobium* (orchid). In: Bajaj YPS (ed.), *Plant Protoplasts and Genetic Engineering VI, Biotechnology in Agriculture and Forestry*, Vol. 34, Springer-Verlag, Berlin, pp. 145–155.
45. Joerbo M. (2001) Advances in the selection of transgenic plants using non-antibiotic marker genes. *Physiologic Planatarum* **111:**269–272.
46. Erikson O, Hertzherg M, Nasholm T. (2004) A conditional marker gene allowing both positive and negative selection in plants. *Nat Biotechnol* **22:**455–458.
47. Wenck A, Hansen G. (2005) Positive selection. *Methods Mol Biol* **286:** 227–236.

48. Lin HJ, Cheng HY, Chen CH, *et al*. (1997) Plant amphipathic proteins delay the hypersensitive response caused by harpin (Pss) and *Pseudomonas syringae* pv. *Syringae*. *Physiol Mol Plant Pathol* **51:**367–376.
49. Tang K, Sun X, Hu Q, *et al*. (2001) Transgenic rice plants expressing the ferredoxin-like protein (*AP1*) from sweet pepper show enhanced resistance to *Xanthomonas oryzae* pv. oryzae. *Plant Sci* **160:**1035–1042.
50. Chang CC, Lin HC, Lin IP, *et al*. (2006) The chloroplast genome of *Phalaenopsis aphrodite* (*Orchidaceae*): Comparative analysis of evolutionary rate with that of grasses and its phylogenetic implications. *Mol Biol Evol* **23:**279–291.
51. Hsu HF, Yang CH. (2002) An orchid (*Oncidium* Gower Ramsey) AP3-like MADS gene regulates floral formation and initiation. *Plant Cell Physiol* **43:**1198–1209.
52. Hsu HF, Huang CH, Chou LT, Yang CH. (2003) Ectopic expression of an orchid (*Oncidium* Gower Ramsey) AGL6-like gene promotes flowering by activation flowering time genes in *Arabidopsis thaliana*. *Plant Cell Physiol* **44:**783–794.
53. Xu Y, Teo LL, Zhou J, *et al*. (2006) Floral organ identity genes in the orchid *Dendrobium crumenatum*. *Plant J* **46:**54–68.
54. Wu XM, Lim SH, Yang WC. (2003) Characterization, expression and phylogenetic study of *R2R3-MYB* genes in orchid. *Plant Mol Biol* **51:**959–972.

Index

www.ingramcontent.com/pod-product-compliance
Lightning Source LLC
Chambersburg PA
CBHW060238220326
41598CB00027B/3974
* 9 7 8 9 8 1 3 2 0 3 4 1 9 *